SpringerBriefs in Statistics

T0184881

For further volumes:
http://www.springer.com/series/8921

Tapas Kumar Chandra

The Borel–Cantelli Lemma

 Springer

Tapas Kumar Chandra
Applied Statistics Division
Indian Statistical Institute
B. T. Road 203
Kolkata 700108, West Bengal
India

ISSN 2191-544X   ISSN 2191-5458   (electronic)
ISBN 978-81-322-0676-7   ISBN 978-81-322-0677-4   (eBook)
DOI 10.1007/978-81-322-0677-4
Springer India Heidelberg New York Dordrecht

Library of Congress Control Number: 2012941419

Printed on acid-free paper

Springer is part of Springer Science+Business Media (www.springer.com)

To
The memory of my teachers
Professor Anil Kumar Bhattacharyya
(1915–1996)
and
Professor Ashok Maitra (1938–2008)

Preface

This monograph is concerned with the theory and applications of the Borel–Cantelli Lemma, hereafter referred to as BCL, although the applications of BCL to the strong laws of large numbers and the laws of the iterated logarithms will not be mentioned here. BCL is indispensible for deriving results on, the almost sure behavior of random variables. Hence almost all textbooks on probability theory contain a discussion on BCL. However, I have tried to include here as an extensive a treatment of BCL as possible. I have attempted to make this monograph self-contained by introducing some standard facts on probability theory in Chap. 1. A special feature of this treatise is a very exhaustive list of research papers and books on BCL; however, if there is any important omission in this regard, it is due to the lack of my knowledge and I sincerely apologize for it.

Attempts have been made to make the discussion lucid, simple, and thorough; the proofs are given in great detail and are completely rigorous. Any advanced undergraduate student learning probability theory will be able to understand a large part of this monograph.

I am grateful to my colleagues Sreela Gangopadhyay and Gour Mohan Saha for their great help. Thanks are also due to Prasanta Kumar Sen for doing an excellent typing.

I learnt the introductory probability from Anil Kumar Bhattacharyya, my teacher at Presidency College, Kolkata, India. Then I learnt the measure theoretic and advanced probability from Ashok Maitra of the Indian Statistical Institute, Kolkata, India. I am indebted to them for my current state of understanding, probability theory. I gratefully dedicate this monograph to the loving memory of these two great teachers and respectable personalities.

March 2012

T. K. Chandra

Contents

Abbreviations

AP	Annals of Probability
AS	Annals of Statistics
AMS	Annals of Mathematical Statistics
IJM	Illinois Journal of Mathematics
JMA	Journal of Multivariate Analysis
JMS	Journal of Mathematical Sciences
MS	Mathematica Scandinavica
PAMS	Proceedings of American Mathematical Sciences
SJAM	Siam Journal of Applied Mathematics
SPL	Statistics and Probability Letters
TAMS	Transactions of American Mathematical Society
TPA	Theory of Probability and its Applications
ZWVG	Zeitschrift für Wahrscheinlichkeitstheorie and Verwandte Gebiete
iff	If and only if
iid	Independent and identically distributed
a.s.	Almost surely
LHS	Left Hand Side
RHS	Right Hand Side
sup	Supremum
inf	Infimum
lim sup	Limit superior
lim inf	Limit inferior
LLN	Law(s) of Large Numbers
SLLN	Strong Law of Large Numbers
BCL	Borel–Cantelli Lemma(s)
NQD	Negative quadrant dependent
$\varnothing$	Empty set
Ω	Underlying set, sample space
$\subset$	Is a subset of
$\in$	Belongs to
$\notin$	Does not belong to

A^c	The complement of the set A
$\cup$	Union
$\cap$	Intersection
Δ	Symmetric difference
$A \backslash B$	$A \cap B^c$
I_A	The indicator function of the set A
A, B etc.	Events
X, Y etc.	Random variables
n, m etc.	Natural numbers
$P(A)$	Probability of the event A
$E(X)$	The expectation of X
N_n	$\sum_{i=1}^{n} I_{A_i}$
s_n	$E(N_n) = \sum_{i=1}^{n} P(A_i)$
N	$\sum_{i=1}^{\infty} I_{A_i}$
S_n	$\sum_{i=1}^{n} X_i$
$\overline{X}_n$	$n^{-1} S_n$
$E(X \ A)$	$E(X \ I_A)$
$\mathrm{var}(X)$	$E(X - E(X))^2$
$\mathrm{cov}(X, Y)$	$E((X - E(X))(Y - E(Y)))$
$X_n \to^P X$	Convergence in probability
$X_n \to X$ a.s.	Almost sure convergence
$N(\mu; \sigma^2)$	Normal distribution with mean μ and variance σ^2
$U(a, b)$	Uniform distribution over (a, b)
Poisson (λ)	Poisson distribution with mean λ
$a_n = o(b_n)$	$a_n/b_n \to 0$
$a_n = 0(b_n)$	$\{|a_n/b_n|\}$ is bounded above
$a_n \sim b_n$	$a_n/b_n \to 1$
$[x]$	The greatest integer $\leq x$
$\langle x \rangle$	The fractional part of x
$\mathbb{Q}$	The set of all rational numbers
$\mathbb{N}$	The set of all natural numbers
$\mathbb{R}$	The set of all reals
$\overline{\mathbb{R}}$	$\mathbb{R} \cup \{\infty, -\infty\}$
a^+	max $\{a, 0\}$
$:=$	Defined as
$\exists$	There exists
$\forall$	For all
$\Rightarrow$	Implies
$\Leftrightarrow$	Implies and is implied by
$\square$	The end of proofs

Chapter 1
Introductory Chapter

1.1 Probability Spaces

Let Ω be a nonempty (abstract) set. Let $\mathcal{A}$ be a σ-**field** of subsets of Ω; i.e., $\mathcal{A}$ is a family of subsets of Ω such that

(a) $\Omega \in \mathcal{A}$;
(b) $A \in \mathcal{A} \Rightarrow A^c \in \mathcal{A}$; and
(c) $A_n \in \mathcal{A} \forall n \geq 1 \Rightarrow \cup_{n=1}^{\infty} A_n \in \mathcal{A}$.

Let P be a **probability measure** defined on $\mathcal{A}$; i.e., P is a set function defined on $\mathcal{A}$ into $\mathbb{R}$ such that

(d) $P(A) \geq 0 \forall A \in \mathcal{A}$;
(e) $P(\Omega) = 1$; and
(f) whenever $\{A_n\}_{n \geq 1}$ is a sequence of pairwise disjoint sets in $\mathcal{A}$, one has

$$\sum_{n=1}^{\infty} P(A_n) \text{ converges to } P\left(\overset{\infty}{\underset{n=1}{\cup}} A_n \right).$$

The triplet $(\Omega, \mathcal{A}, P)$ is called a **probability space**. It has the following interpretation. We have a random experiment in mind (an experiment is called **random** if the set of all possible outcomes of it is known but it is impossible to foretel which one of these outcomes will occur in case the experiment is performed once). Then Ω will stand for its **sample space**, i.e., the set of all possible outcomes of the experiment; the family $\mathcal{A}$ will stand for the set of all possible **events**; and P will denote the chance mechanism governing the occurrence of the random outcomes (or the events). By an **event** A, we shall mean that $A \in \mathcal{A}$. This axiomatization of a probability space is due to Kolmogorov (1933).

We shall be concerned with some of the basic properties of the probability space. First, $\mathcal{A}$ is closed under *finitely many or countably many* set operations. Second,

T. K. Chandra, *The Borel–Cantelli Lemma*, SpringerBriefs in Statistics, DOI: 10.1007/978-81-322-0677-4_1, © The Author(s) 2012

$P(\emptyset) = 0$ and P is finitely additive. Also,

(g) $P(A \backslash B) = P(A) - P(B)$ if $B \subset A$, and $B, A \in \mathcal{A}$;

(h) (**the truncation inequality**) $P(A) \leq P(A \cap B) + P(B^c)$ if $A, B \in \mathcal{A}$; and so

$$|P(A) - P(B)| \leq P(A \triangle B) \text{ if } A, B \in \mathcal{A},$$

$\triangle$ being the symmetric difference operator;

(i) $P(B) \leq P(A)$ if $B \subset A$, and $B, A \in \mathcal{A}$;

(j) $P(A^c) = 1 - P(A)$ if $A \in \mathcal{A}$;

(k) (**Boole's inequality**) if $A_n \in \mathcal{A} \forall n \geq 1$, then

$$P\left(\bigcup_{n=1}^{\infty} A_n\right) \leq \sum_{n=1}^{\infty} P(A_n);$$

in particular, if $A_1, \ldots, A_n \in \mathcal{A}$, then

$$P\left(\bigcup_{i=1}^{n} A_i\right) \leq \sum_{i=1}^{n} P(A_i);$$

(l) if $A_n \subset A_{n+1}$ and $A_n \in \mathcal{A} \forall n \geq 1$, then

$$P(A_n) \to P\left(\bigcup_{n=1}^{\infty} A_n\right);$$

(m) if $A_{n+1} \subset A_n$ and $A_n \in \mathcal{A} \forall n \geq 1$, then

$$P(A_n) \to P\left(\bigcap_{n=1}^{\infty} A_n\right);$$

(n) $P(A \cup B) = P(A) + P(B) - P(A \cap B)$ if $A, B \in \mathcal{A}$.

The facts (l) and (m) are known as the **continuity properties** of the set function P. If $A_n \subset A_{n+1} \forall n \geq 1$, one says that $\{A_n\}_{n \geq 1}$ is **increasing**, and writes $A_n \uparrow A$ where $A = \cup_{n=1}^{\infty} A_n$; if $A_{n+1} \subset A_n \forall n \geq 1$, one says that $\{A_n\}_{n \geq 1}$ is **decreasing**, and writes $A_n \downarrow A$ where $A = \cap_{n=1}^{\infty} A_n$.

The sequence $\{A_n\}_{n \geq 1}$ is called **monotone** if it is either increasing or decreasing. Thus (l), (m) together say that

$$\{A_n\}_{n \geq 1} \text{ is monotone}, A_n \in \mathcal{A} \forall n \geq 1 \quad and \quad A_n \to A \Rightarrow P(A_n) \to P(A).$$

The proofs of (g)–(j) and (n) are elementary. To prove (k), put

$$B_n = A_n \backslash \left(\bigcup_{i=0}^{n-1} A_i\right), \quad n \geq 1$$

where $A_0 = \emptyset$; the sets B_n are obtained from the A_n *successively by the process of disjointification*. Obviously, $B_n \in \mathcal{A} \ \forall n \geq 1$, $B_n \subset A_n$ and $\cup_{n=1}^{\infty} A_n = \cup_{n=1}^{\infty} B_n$; furthermore, $B_n \cap B_m = \emptyset$ if $n \neq m$, for, $n \leq m - 1 \Rightarrow B_m \subset A_n^c$. Thus

$$P\left(\bigcup_{n=1}^{\infty} A_n\right) = P\left(\bigcup_{n=1}^{\infty} B_n\right) = \sum_{n=1}^{\infty} P(B_n) \leq \sum_{n=1}^{\infty} P(A_n).$$

The fact (l) can be established thus: Define $B_n = A_n \backslash A_{n-1} \ \forall n \geq 1$ where $A_0 = \emptyset$. Then $B_n \in \mathcal{A} \ \forall n \geq 1$, the sets B_n are pairwise disjoint, $\cup_{n=1}^{\infty} B_n = \cup_{n=1}^{\infty} A_n$ and $\cup_{i=1}^{n} B_i = A_n$, $\forall n \geq 1$. Thus

$$P\left(\bigcup_{n=1}^{\infty} A_n\right) = P\left(\bigcup_{n=1}^{\infty} B_n\right) = \sum_{n=1}^{\infty} P(B_n)$$

$$= \lim_{n \to \infty} \sum_{i=1}^{n} P(B_i) = \lim_{n \to \infty} P\left(\bigcup_{i=1}^{n} B_i\right) = \lim_{n \to \infty} P(A_n).$$

The fact (m) follows from (l) and (j) as can be seen thus: Put $B_n = A_n^c, n \geq 1$. Then $\{B_n\}$ is increasing so that, by (l),

$$P(B_n) \to P\left(\bigcup_{n=1}^{\infty} B_n\right)$$

and hence $1 - P(A_n) \to 1 - P\left(\cap_{n=1}^{\infty} A_n\right)$ by (j) and de Morgan's law.

We note that if $A_n \in \mathcal{A} \ \forall n \geq 1$, then

$$P\left(\bigcup_{n=1}^{\infty} A_n\right) = 0 \Leftrightarrow P(A_n) = 0 \ \forall n \geq 1; \text{ and}$$

$$P\left(\bigcap_{n=1}^{\infty} A_n\right) = 1 \Leftrightarrow P(A_n) = 1 \ \forall n \geq 1.$$

Indicator Functions

There is a very nice, elegant, and useful duality between sets and functions in the context of an abstract set, which emphasizes their algebraic properties.

Let $A \subset \Omega$. By the indicator function, I_A, of A, we shall mean a function from Ω into $\{0, 1\}$ such that

$$I_A(\omega) = \begin{cases} 1 \text{ if } \omega \in A; \\ 0 \text{ if } \omega \notin A. \end{cases}$$

Many useful properties of indicator functions are known. We mention the following one only: If $T \neq \emptyset$, then with $A = \cup_{t \in T} A_t$ and $B = \cap_{t \in T} A_t$

$$I_A = \sup_{t \in T} I_{A_t}, \quad I_B = \inf_{t \in T} I_{A_t}.$$

In a probability space $(\Omega, \mathcal{A}, P)$, for a subset A of Ω one has

$$I_A \text{ is a random variable} \Leftrightarrow A \text{ is an event.}$$

At this stage it is recommended that the reader should go through Sects. 15, 16, 20, and 21 of Billingsley (1995).

Lemma 1.1.1 *Let X be a non-negative random variable.*

(a) **(The Expectation Identity)** *If $\sum_{n=0}^{\infty} P(X = n) = 1$, then*

$$E(X) = \sum_{n=1}^{\infty} P(X \geq n).$$

In general,

$$E([X]) = \sum_{n=1}^{\infty} P(X \geq n),$$

where $[X]$ is the integer part of X.

(b) **(The Expectation Inequality)**

$$\sum_{n=1}^{\infty} P(X \geq n) \leq E(X) \leq 1 + \sum_{n=1}^{\infty} P(X \geq n).$$

In particular, $E(|X|) < \infty \Leftrightarrow \sum_{n=1}^{\infty} P(|X| \geq n) < \infty.$

Proof (a) Note that

$$E(X) = \sum_{n=1}^{\infty} n \, P(X = n) = \sum_{n=1}^{\infty} \left(\sum_{i=1}^{n} 1 \right) P(X = n)$$

$$= \sum_{i=1}^{\infty} \sum_{n=i}^{\infty} P(X = n) = \sum_{i=1}^{\infty} P(X \geq i),$$

and

$$E([X]) = \sum_{n=1}^{\infty} P([X] \geq n) = \sum_{n=1}^{\infty} P(X \geq n).$$

(b) This is immediate, since $[X] \leq X \leq 1 + [X]$. $\qquad\qquad\square$

Lemma 1.1.2 *The events* $A_1, \ldots, A_n$ $(n \geq 2)$ *are independent iff*

$$P\left(\bigcap_{i=1}^{n} B_i\right) = \prod_{i=1}^{n} P(B_i)$$

for all choices of the B_i *such that* $B_i = A_i$ *or* A_i^c *for* $1 \leq i \leq n$.

Lemma 1.1.3 *A random variable* X *is degenerate iff for each real* x, $P(X \leq x)$ *is 0 or 1.*

1.2 Lim Sup and Lim Inf of a Sequence of Sets

We shall introduce the important notions of "lim sup" and "lim inf" of a sequence of subsets of an abstract set Ω. This involves set-theoretic operations only, although we shall use these notions in various contexts freely.

Let Ω be an abstract set.

Definition 1.2.1 Let $\{A_n\}_{n \geq 1}$ be a sequence of subsets of the set Ω.

(a) We define $\limsup_{n \to \infty} A_n = \bigcap_{n=1}^{\infty} \bigcup_{m=n}^{\infty} A_m$.
(b) We define $\liminf_{n \to \infty} A_n = \bigcup_{n=1}^{\infty} \bigcap_{m=n}^{\infty} A_m$.
(c) We say that $\lim_{n \to \infty} A_n$ exists if $\limsup_{n \to \infty} A_n = \liminf_{n \to \infty} A_n$; in this case we write $\lim_{n \to \infty} A_n = \limsup_{n \to \infty} A_n$.

When there is no chance of confusion, we replace

$$\limsup_{n \to \infty} A_n, \ \liminf_{n \to \infty} A_n \text{ and } \lim_{n \to \infty} A_n$$

by, respectively,

$$\limsup A_n, \liminf A_n \text{ and } \lim A_n.$$

We shall often write $A_n \to A$ in case $\lim_{n \to \infty} A_n = A$.

To bring the analogy with the "lim sup" and "lim inf" of real sequences, recall that

$$\limsup a_n = \inf_{n \geq 1} \sup_{m \geq n} a_m, \ \liminf a_n = \sup_{n \geq 1} \inf_{m \geq n} a_m.$$

(We shall be concerned below with only *bounded* sequences $\{a_n\}_{n \geq 1}$.)
The relationship between the above definitions is given below:

$$I_{\limsup A_n} = \limsup I_{A_n}, \ I_{\liminf A_n} = \liminf I_{A_n}.$$

and

$$A_n \to A \Leftrightarrow I_{A_n} \to I_A.$$

Furthermore, if $\{A_n\}_{n\geq 1}$ is increasing, then

$$\limsup A_n = \liminf A_n = \bigcup_{n=1}^{\infty} A_n,$$

while if $\{A_n\}_{n\geq 1}$ is decreasing, then

$$\limsup A_n = \liminf A_n = \bigcap_{n=1}^{\infty} A_n.$$

In general,

$$\bigcup_{m=n}^{\infty} A_m \downarrow \limsup A_n, \quad \bigcap_{m=n}^{\infty} A_m \uparrow \liminf A_n. \quad (\star)$$

By de Morgan's laws, it is obvious that

$$(\liminf A_n)^c = \limsup A_n^c, \ (\limsup A_n)^c = \liminf A_n^c.$$

Thus, in a sense, one of the two notions suffices.

Finally, note that

(A) $\omega \in \limsup A_n$
  $\Leftrightarrow$ for each $n \geq 1$, there is an integer $m \geq n$ such that $\omega \in A_m$
  $\Leftrightarrow$ $\omega \in A_n$ for infinitely many values of n;
(B) $\omega \in \liminf A_n$
  $\Leftrightarrow$ there is an integer $n \geq 1$ such that $\omega \in A_m \ \forall m \geq n$
  $\Leftrightarrow$ $\omega \in A_n$ for all sufficiently large n (or, simply, $\omega \in A_n$ eventually).

Incidentally, the above alternative descriptions of $\limsup A_n$ and $\liminf A_n$ show that

(C) these two sets are free from any particular enumeration of the sets A_n as a sequence;
(D) these two sets remain unaffected, even if we change the sets A_n for only finitely many values of n;
(E) $\liminf A_n \subset \limsup A_n$; and
(F) if $n_1 < n_2 < \cdots$, then $\limsup A_{n_k} \subset \limsup A_n$.

In the language of probability,

$$\limsup A_n = [A_n i.o.(n)] = [A_n \ i.o.] \text{ if there is no confusion;}$$
$$i.o. = \text{ infinitely often;}$$
$$\liminf A_n = [A_n \text{ eventually } (n)] = [A_n \text{ eventually }] \text{ if there is no confusion.}$$

If now there is a σ-field $\mathcal{A}$ of subsets of Ω, and $A_n \in \mathcal{A} \ \forall n \geq 1$, then $\limsup A_n \in \mathcal{A}$ and $\liminf A_n \in \mathcal{A}$ (and $A_n \to A \Rightarrow A \in \mathcal{A}$).

Lemma 1.2.1 *Let* $(\Omega, \mathcal{A}, P)$ *be a probability space, and* $A_n \in \mathcal{A} \ \forall n \geq 1$. *Then*
(a)

$$P(\liminf A_n) \leq \liminf P(A_n) \leq \limsup P(A_n) \leq P(\limsup A_n); \quad (1.2.1)$$

$$in\ particular,\ A_n \to A \Rightarrow P(A_n) \to P(A_n)(and\ P(A_n \triangle A) \to 0); \quad (1.2.2)$$

$$also,\ P(\limsup A_n) = 0 \Rightarrow P(A_n) \to 0;\ and \quad (1.2.3)$$

$$\liminf P(A_n) = 0 \Rightarrow P(\liminf A_n) = 0; \quad (1.2.4)$$

(b) *if* $A \in \mathcal{A}$,

$$P((\limsup A_n) \cap A^c) = 0 \Rightarrow P(A_n \cap A^c) \to 0,$$
$$P((\liminf A_n)^c \cap A) = 0 \Rightarrow P(A_n^c \cap A) \to 0, \quad (1.2.5)$$

(thus (1.2.5) is an extension of (1.2.2));
(c) $P(A_n) \geq \delta > 0 \ \forall n \geq 1$ *where* δ *is free from* $n \Rightarrow \limsup A_n \neq \emptyset$;
(d) $P(\limsup A_n) = \lim P\left(\bigcup_{m=n}^{\infty} A_m\right)$, $P(\liminf A_n) = \lim_{n \to \infty} P\left(\bigcap_{m=n}^{\infty} A_m\right)$.

Proof (a)   This follows from earlier results (see ($\star$) of p. 6 and (m) of Sect. 1.1) in the following way:

$$P(\limsup A_n) = \lim_{n \to \infty} P\left(\bigcup_{m=n}^{\infty} A_m\right) \geq \limsup P(A_n).$$

Also,

$$P(\liminf A_n) = 1 - P(\limsup A_n^c)$$
$$\leq 1 - \limsup P(A_n^c) \text{ by what has been proved}$$
$$= \liminf P(A_n).$$

The remaining parts of (a) are now easy to prove.
(b)   Note that

$$0 = P((\limsup A_n) \cap A^c) = P(\limsup(A_n \cap A^c))$$

and hence $P(A_n \cap A^c) \to 0$ by (1.2.2). Now replacing A_n by A_n^c for each $n \geq 1$ and A by A^c, we get

$$P((\limsup A_n^c) \cap A) = 0 \Rightarrow P(A_n^c \cap A) \to 0.$$

We now get (1.2.5).

(c) This is immediate since

$$P(\limsup A_n) \geq \limsup P(A_n) \geq \delta > 0.$$

(d) It follows from (l) and (m) of Sect. 1.1; see the proof of (a).   □

We next state a few facts for later uses:

$$A_n \subset B_n \forall n \geq 1 \Rightarrow \limsup A_n \subset \limsup B_n, \ \liminf A_n \subset \liminf B_n, \qquad (1.2.6)$$
$$(\limsup A_n) \cup (\limsup B_n) = \limsup(A_n \cup B_n), \qquad (1.2.7)$$
$$(\liminf A_n) \cap (\liminf B_n) = \liminf(A_n \cap B_n), \qquad (1.2.8)$$
$$(\limsup A_n) \cap (\limsup B_n) \supset \limsup(A_n \cap B_n), \qquad (1.2.9)$$
$$(\liminf A_n) \cup (\liminf B_n) \subset \liminf(A_n \cup B_n), \qquad (1.2.10)$$

(The two inclusions may be **strict**.)

$$A_n \to A, \ B_n \to B \Rightarrow A_n \cup B_n \to A \cup B, \ A_n \cap B_n \to A \cap B. \qquad (1.2.11)$$

Lemma 1.2.2 *The following is true for any sequence of sets* $\{A_n\}_{n \geq 1}$.

$$(\limsup A_n) \cap (\limsup A_n^c) = \limsup(A_n \cap A_{n+1}^c)$$
$$= \limsup(A_n^c \cap A_{n+1}).$$

Proof In view of (1.2.6), it suffices to show that

$$(\limsup A_n) \cap (\limsup A_n^c) \subset \limsup(A_n \cap A_{n+1}^c), \qquad (1.2.12)$$
$$(\limsup A_n) \cap (\limsup A_n^c) \subset \limsup(A_n^c \cap A_{n+1}). \qquad (1.2.13)$$

The inclusion in (1.2.13) follows from that of (1.2.12) by replacing A_n by A_n^c for each $n \geq 1$. It remains to prove (1.2.12). To this end, let $\omega \in$ LHS of (1.2.12). Fix an integer $n \geq 1$. There is an integer $m \geq n$ and that $\omega \in A_m$ (as $\omega \in \limsup_{i \to \infty} A_i$). Put

$$k = \inf\{j > m : \omega \in A_j^c\}.$$

As $\omega \in \limsup_{i \to \infty} A_i^c$, there is an integer $j \geq m+1$ such that $\omega \in A_j^c$; thus k is finite and $k \geq m+1 \geq 2$. Now observe that

$$\omega \in A_k^c \cap A_{k-1},$$

by considering the following two cases:

Case 1 $k = m+1$.

Then $\omega \in A_k^c \cap A_m = A_k^c \cap A_{k-1}$.

Case 2 $k \geq m + 2$.

Then $k - 1 \geq m + 1$ so that $\omega \notin A_{k-1}^c$. Thus $\omega \in A_k^c \cap A_{k-1}$.
As $k \geq m + 1 \geq n + 1 \geq n$, we must have $\omega \in$ RHS of (1.2.12).

Example 1.2.1 Show that if $P(\liminf A_n) = P(\limsup B_n) = 1$, then

$$P(\limsup(A_n \cap B_n)) = 1.$$

Solution: As $P((\liminf A_n) \cap (\limsup B_n)) = 1$, it suffices to show that

$$(\liminf A_n) \cap (\limsup B_n) \subset \limsup(A_n \cap B_n). \qquad (1.2.14)$$

But if ω is in the left side, then there is an integer $k \geq 1$ such that $\omega \in A_n$ $\forall n \geq k$, and $\omega \in B_n$ for $n = n_1, n_2, \ldots$ where $n_1 < n_2 < \cdots$ Thus $\omega \in A_n \cap B_n$ for $n = n_j, n_{j+1}, \ldots$ where $j \geq 1$ is such that $n_j \geq k$.

Example 1.2.2 Let $\{A_n\}_{n \geq 1}$ and $\{B_n\}_{n \geq 1}$ be two sequences of events such that there is an integer $m \geq 1$ satisfying A_n is independent of $\{B_n, B_{n+1}, \ldots\}$ $\forall n \geq m$. Show that

$$P(\limsup(A_n \cap B_n)) \geq \liminf P(A_n)P(\limsup B_n).$$

Solution: Fix an integer $n \geq m$. Note that for each $i \geq 1$,

$$P\left(\bigcup_{k=n}^{\infty} (A_k \cap B_k)\right) \geq P\left(\bigcup_{k=n}^{n+i} (A_k \cap B_k)\right)$$

$$= \sum_{k=n}^{n+i-1} P\left((A_k \cap B_k) \cap \bigcap_{j=k+1}^{n+i} (A_j \cap B_j)^c\right)$$

$$+ P(A_{n+i} \cap B_{n+i})$$

by a standard disjointificaion of the sets $(A_{n+i} \cap B_{n+i}), \ldots, (A_n \cap B_n)$; see, e.g., the proof of Boole's inequality in p. 2.

Thus for each $i \geq 1$,

$$P\left(\bigcup_{k=n}^{\infty} (A_k \cap B_k)\right) \geq \sum_{k=n}^{n+i-1} P\left((A_k \cap B_k) \cap \bigcap_{j=k+1}^{n+i} B_j^c\right) + P(A_{n+i} \cap B_{n+i})$$

$$= \sum_{k=n}^{n+i-1} P(A_k)P\left(B_k \cap \bigcap_{j=k+1}^{n+i} B_j^c\right) + P(A_{n+i})P(B_{n+i})$$

$$\geq \inf_{k \geq n} P(A_k)\left[\sum_{k=n}^{n+i-1} P\left(B_k \cap \bigcap_{j=k+1}^{n+i} B_j^c\right) + P(B_{n+i})\right]$$

$$= \inf_{k \geq n} P(A_k)P\left(\bigcup_{k=n}^{n+i} B_k\right).$$

Letting $i \to \infty$, we get

$$P \left(\bigcup_{k=n}^{\infty} (A_k \cap B_k) \right) \geq \inf_{k \geq n} P(A_k) P \left(\bigcup_{k=n}^{\infty} B_k \right)$$

$$\geq \inf_{k \geq n} P(A_k) P (\limsup B_n).$$

Now letting $n \to \infty$, we get the desired result.

The above example is closely related to a result of Feller and Chung; see, in this connection, pp. 69–70 of Chow and Teicher (1997).

To illustrate how the use of "lim sup" and "lim inf" makes complicated statements very transparent, we first recall the following well-known definitions.

Let each X_n and X be random variables defined on $(\Omega, \mathcal{A}, P)$.

Definition 1.2.2 We say that $\{X_n\}_{n \geq 1}$ **converges almost surely to** X, written $X_n \to X$ a.s. [P], if there is an event $A \in \mathcal{A}$ such that $P(A) = 1$ and

$$A \subset [X_n \to X].$$

Since each of the X_n and X are measurable functions, the set $[X_n \to X]$ lies in $\mathcal{A}$ (see (1.2.16) on p. 11 below); thus

$$X_n \to X \ a.s. [P] \iff P(X_n \to X) = 1.$$

Definition 1.2.3 We say that $\{X_n\}_{n \geq 1}$ **converges in probability to** X, written $X_n \to^P X$, if for each $\epsilon > 0$

$$P(|X_n - X| > \epsilon) \to 0. \tag{1.2.15}$$

Let $0 < \delta < \infty$; then

$$X_n \to^P X \iff P(|X_n - X| > \epsilon) \to 0 \ \forall \ 0 < \epsilon < \delta.$$

In other words, in (1.2.15) one may require ϵ to be sufficiently small.

At this stage, it is instructive to supply the details of the following examples.

Example 1.2.3 Let $\{X_n\}_{n \geq 1}$ satisfy

$$P(|X_n| > x) \leq P(|Y| > x) \ \forall \ x > 0, n \geq 1$$

and $E(|Y|^p) < \infty$ for some $p > 0$. If $Y_n = \max(|X_1|, \ldots, |X_n|)$ for $n \geq 1$, then show that $n^{-1/p} Y_n \to^P 0$.

Example 1.2.4 Assume that for each $i = 1, 2, \ldots, \{X_{1,i}, \ldots, X_{n_i,i}\}$ be independent. Show that

$$\max_{1 \le j \le n_k} |X_{j,k}| \to^P 0 \text{ iff } \sum_{j=1}^{n_k} P(|X_{j,k}| > \epsilon) \to 0 \,\forall\, \epsilon > 0.$$

Lemma 1.2.3 *Let* $X_n \to X$ *a.s.[P]. Then*

(a) $P(|X_n - X| \ge \epsilon \text{ i.o.}) = 0$ *for each* $\epsilon > 0$;
(b) $X_n \to^P X$.

Proof (a) Let $\epsilon > 0$, and put $A_n = [|X_n - X| \ge \epsilon]$, $n \ge 1$. Then $A_n \in \mathcal{A} \,\forall\, n \ge 1$, and obviously

$$\limsup A_n \subset [X_n \not\to X].$$

As $X_n \to X$ a.s. [P], $P(X_n \not\to X) = 0$ so that $P(\limsup A_n) = 0$.
(b) Let $\epsilon > 0$, and define A_n as above. Then $P(\limsup A_n) = 0$ by (a), and so $P(A_n) \to 0$ by (1.2.3). This means that $X_n \to^P X$. $\qquad\square$

Theorem 1.2.1 *Let* $\epsilon_m \to 0+$ *as* $m \to \infty$. *Then the following are equivalent:*

(a) $X_n \to X$ *a.s. [P].*
(b) $P(|X_n - X| > \epsilon \text{ i.o.}) = 0 \,\forall\, \epsilon > 0$. *Or,* $P(|X_n - X| \ge \epsilon \text{ i.o.}) = 0 \,\forall\, \epsilon > 0$.
(c) $P(|X_n - X| > \epsilon_m \text{ i.o. } (n)) = 0 \,\forall\, m \ge 1$.
(d) $P\left(\bigcap_{m=1}^{\infty} \bigcup_{k=1}^{\infty} \bigcap_{n=k}^{\infty} [|X_n - X| \le \epsilon_m]\right) = 1$.

Proof It is clear that (a) and (d) are equivalent, since

$$a_n \to a \text{ as } n \to \infty$$

iff for each $m \ge 1$, there exists an integer $k \ge 1$ such that

$$|a_n - a| \le \epsilon_m \,\forall\, n \ge k,$$

which implies that

$$[X_n \to X] = \bigcap_{m=1}^{\infty} \bigcup_{k=1}^{\infty} \bigcap_{n=k}^{\infty} [|X_n - X| \le \epsilon_m]. \tag{1.2.16}$$

[Note that (1.2.16) is true for **any sequence of functions from** Ω **into** $\mathbb{R}$.]

(a) $\Rightarrow$ (b): See Lemma 1.2.3 (a).
(b) $\Rightarrow$ (c): Trivially true.
(c) $\Rightarrow$ (d): Let $A_m = [|X_n - X| > \epsilon_m \text{ i.o.}(n)]$, $m \ge 1$. Then

$P(A_m) = 0 \,\forall\, m \ge 1$ by (c). Hence $P\left(\bigcup_{m=1}^{\infty} A_m\right) = 0$; i.e.,

$$1 = P\left(\bigcap_{m=1}^{\infty} A_m^c\right) = P\left(\bigcap_{m=1}^{\infty} \bigcup_{k=1}^{\infty} \bigcap_{n=k}^{m} [|X_n - X| \le \epsilon_m]\right). \qquad\square$$

There is an analogue of (1.2.16) for "the Cauchy condition" for $\{X_n\}_{n\geq 1}$: Let $\{X_n\}_{n\geq 1}$ be **any sequence of funcions from Ω into $\mathbb{R}$**. Then

$$[\{X_n\} \text{ is Cauchy }] = \bigcap_{m=1}^{\infty} \bigcup_{n=1}^{\infty} \bigcap_{k=n+1}^{\infty} [|X_n - X_k| \leq \epsilon_m]$$

$$= \bigcap_{m=1}^{\infty} \bigcup_{n=1}^{\infty} \bigcap_{k=n+1}^{\infty} [\max_{n < i \leq k} |X_n - X_i| \leq \epsilon_m]$$

where $\epsilon_m \to 0+$ is a fixed real sequence. **Now assume, furthermore, that $\epsilon_m \downarrow 0$, and the X_n are measurable**; then

$$P(\{X_n\} \text{ is Cauchy }) = \lim_{m \to \infty} P\left(\bigcup_{n=1}^{\infty} \bigcap_{k=n+1}^{\infty} [|X_n - X_k| \leq \epsilon_m]\right)$$

$$= \lim_{m \to \infty} \lim_{n \to \infty} \lim_{k \to \infty} P\left(\max_{n < i \leq k} |X_n - X_i| \leq \epsilon_m\right).$$

These facts are straightforward to verify. Since for a real sequence $\{a_n\}_{n\geq 1}$, it is known that $\lim a_n$ exists and is finite $\Leftrightarrow \{a_n\}$ is Cauchy, we have

$$\exists X \text{ such that } X_n \to X a.s. \Leftrightarrow P(\{X_n\} \text{ is Cauchy}) = 1.$$

[We have thus obtained a criterion for the *a.s.* convergence of $\{X_n\}$ in terms of the finite dimensional distributions of $\{X_n\}$.]

We conclude this section with some applications of the above results.

Example 1.2.5 Let $X_n \to X$ a.s. and the distribution of $(X_1, \ldots, X_n)$ be identical with that of $(Y_1, \ldots, Y_n)$ for each $n \geq 1$ (the Y_n are defined possibly on a second probability space). Show that $\exists$ a random variable Y such that $Y_n \to Y$ a.s. and $X \overset{d}{=} Y$.

Solution: As observed earlier,

$$X_n \to X \text{ a.s. } \Leftrightarrow \lim_{m \to \infty} \lim_{n \to \infty} \lim_{k \to \infty} P\left(\max_{n < i \leq k} |X_n - X_i| \leq \frac{1}{m}\right) = 1.$$

As the distributions of $(X_1, \ldots, X_k)$ and $(Y_1, \ldots, Y_k)$ are same, so are those of $(X_n, \ldots, X_k)$ and $(Y_n, \ldots, Y_k)$ where $1 \leq n < k < \infty$. Therefore, for each $1 \leq n < k < \infty$ and $m \geq 1$

$$P\left(\max_{n < i \leq k} |X_n - X_i| \leq \frac{1}{m}\right) = P\left(\max_{n < i \leq k} |Y_n - Y_i| \leq \frac{1}{m}\right).$$

This implies

$$\lim_{m \to \infty} \lim_{n \to \infty} \lim_{k \to \infty} P\left(\max_{n < i \leq k} |Y_n - Y_i| \leq 1/m\right) = 1$$

i.e., $\exists$ a random variable Y such that $Y_n \to Y$ a.s.

Example 1.2.6 (Shuster (1970)) If $\{A_n\}_{n\geq 1}$ is such that

$$P(A) > 0 \Rightarrow \sum_{n=1}^{\infty} P(A_n \cap A) = \infty,$$

then $P(\limsup A_n) = 1$.

[Note that $\{A_n\}_{n\geq 1}$ need not be independent; compare with the second Borel–Cantelli lemma (of the next section) where the independence of the A_n is needed.]

Solution: It suffices to show that

$$P\left(\bigcup_{k=m}^{\infty} A_k\right) = 1 \ \forall \ m \geq 1. \tag{1.2.17}$$

Suppose this is false. Then $\exists$ an integer $m \geq 2$ such that $P\left(\bigcup_{k=m}^{\infty} A_k\right) < 1$. Put $A = \bigcap_{k=m}^{\infty} A_k^c$. Then $P(A) > 0$, but

$$\sum_{n=1}^{\infty} P(A_n \cap A) = \sum_{n=1}^{m-1} P(A_n \cap A) < \infty.$$

This contradicts the given condition.

Remark 1.2.1 Let $\sum P(A_n) = \infty$ and suppose that for each event B

$$\liminf \sum_{i=1}^{n} (P(A_i \cap B) - P(A_i)P(B)) \neq -\infty.$$

Then $P(\limsup A_n) = 1$. This is immediate from Example 1.2.6, since one has with $d_n = \sum_{i=1}^{n} (P(A_i \cap B) - P(A_i)P(B))$

$$P(B) > 0 \Rightarrow \sum_{i=1}^{n} P(A_i \cap B) = \left(\sum_{i=1}^{n} P(A_i)\right) P(B) + d_n \to \infty.$$

Example 1.2.7 Let $\{A_n\}_{n\geq 1}$ be a sequence of independent events such that $P(A_n) < 1 \ \forall \ n \geq 1$. Show that

$$P(\limsup A_n) = 1 \Leftrightarrow P\left(\bigcup_{n=1}^{\infty} A_n\right) = 1.$$

Solution: We shall show the implicaion $\Leftarrow$. So let $P\left(\bigcup_{n=1}^{\infty} A_n\right) = 1$. It is enough to show (1.2.17). The result is true for $m = 1$. Let it be true for $m(\geq 1)$. We now show that $P\left(\bigcup_{n=m+1}^{\infty} A_n\right) = 1$.

To this end, note that

$$1 = P\left(\overset{\infty}{\underset{n=m}{\cup}} A_n\right) = \lim_{k\to\infty} P\left(\overset{m+k}{\underset{n=m}{\cup}} A_n\right)$$

$$= \lim_{k\to\infty}\left[P(A_m) + P\left(\overset{m+k}{\underset{n=m+1}{\cup}} A_n\right) - P(A_m)\, P\left(\overset{m+k}{\underset{n=m+1}{\cup}} A_n\right)\right]$$

by the independence $\{A_m, A_{m+1}, \ldots, A_{m+k}\}$. Thus

$$1 = P(A_m) + P\left(\overset{\infty}{\underset{n=m+1}{\cup}} A_n\right) - P(A_m)\, P\left(\overset{\infty}{\underset{n=m+1}{\cup}} A_n\right)$$

or,

$$(1 - P(A_m))\left(1 - P\left(\overset{\infty}{\underset{n=m+1}{\cup}} A_n\right)\right) = 0.$$

As $P(A_m) < 1$, we must have $P\left(\cup_{n=m+1}^{\infty} A_n\right) = 1$.

Example 1.2.8 (a) Suppose that whenever $r < s$,

$$P([X_n < r \; i.o.] \cap [X_n > s \; i.o.]) = 0.$$

Then $P(\lim X_n \text{ exists in } \bar{\mathbb{R}}) = 1$.
(b) Suppose that for some $r < s$,

$$P(X_n < r \; i.o.) = 1, \; P(X_n > s \; i.o.) = 1.$$

Then $P(\lim X_n \text{ exists in } \bar{\mathbb{R}}) = 0$.

Solution:

(a) Let $\mathbb{Q}$ be the set of all rationals. For $r < s$, let

$$N_{r,s} = [X_n < r \; i.o.] \cap [X_n > s \; i.o.].$$

Observe now that $P\left(\cup_{r,s\in\mathbb{Q}} N_{r,s}\right) = 0$ by the given condition, and that

$$[\liminf X_n < \limsup X_n] \subset \underset{r,s\in\mathbb{Q}}{\cup} N_{r,s}.$$

(b) It suffices to observe that the given conditions imply that

$$P(\liminf X_n \le r) = 1, \; P(\limsup X_n \ge s) = 1,$$

and that

$$[\liminf X_n < \limsup X_n] \subset [\liminf X_n \le r < s \le \limsup X_n].$$

1.3 The Borel–Cantelli Lemma

We now turn to the celebrated Borel–Cantelli lemma, the central theme of this monograph. We first introduce a useful definition: by an event A in a probability space $(\Omega, \mathcal{A}, P)$, we mean that $A \in \mathcal{A}$.

Theorem 1.3.1 (The Borel–Cantelli Lemma). *Let $(\Omega, \mathcal{A}, P)$ be a probability space, and $\{A_n\}_{n \geq 1}$ a sequence of events.*

(a) If $\sum_{n=1}^{\infty} P(A_n)$ converges, then $P(\limsup A_n) = 0$.
(b) If the events A_n are independent and $\sum_{n=1}^{\infty} P(A_n)$ diverges (i.e., $\sum_{n=1}^{\infty} P(A_n) = \infty$), then $P(\limsup A_n) = 1$.

Proof (a) First, note that $\limsup A_n \subset \cup_{m=k}^{\infty} A_m \; \forall \, k \geq 1$. Thus for each integer $k \geq 1$,

$$P(\limsup A_n) \leq P\left(\bigcup_{m=k}^{\infty} A_m\right) \leq \sum_{m=k}^{\infty} P(A_m) \qquad (1.3.1)$$

by Boole's inequality. As $\sum_{n=1}^{\infty} P(A_n)$ converges, the tails $\sum_{m=k}^{\infty} P(A_m) \to 0$ as $k \to \infty$. Letting $k \to \infty$ in (1.3.1), we get the desired result.

(b) We show that $P((\limsup A_n)^c) = 0$. But by definition,

$$(\limsup A_n)^c = \bigcup_{n=1}^{\infty} \bigcap_{m=n}^{\infty} A_m^c,$$

(use the de Morgan laws). It, therefore, suffices to show that for each $n \geq 1$,

$$P\left(\bigcap_{m=n}^{\infty} A_m^c\right) = 0. \qquad (1.3.2)$$

Fix such an $n \geq 1$. Since $1 + x \leq \exp(x)$ for each real x, we have for each $j \geq 1$,

$$P\left(\bigcap_{m=n}^{\infty} A_m^c\right) \leq P\left(\bigcap_{m=n}^{n+j} A_m^c\right)$$

$$= \prod_{m=n}^{n+j} (1 - P(A_m)) \text{ by the independence of } A_n, \ldots, A_{n+j}$$

$$\leq \exp\left(-\sum_{m=n}^{n+j} P(A_m)\right).$$

(Note that $A_n, \ldots, A_{n+j}$ are independent and use Lemma 1.1.2.) As $\sum_{n=1}^{\infty} P(A_n)$ diverges, $\sum_{m=n}^{\infty} P(A_m) = \infty$ so that $\sum_{m=n}^{n+j} P(A_m) \to \infty$ as $j \to \infty$.

Since $\lim_{x \to \infty} \exp(-x) = 0$, we get (1.3.2) by letting $j \to \infty$ in the above inequality.   □

The parts (a) and (b) together are known as the Borel–Cantelli lemma, in short BCL, the former the "convergence part" and the latter "the divergence part". However, it will be convenient to refer to (a) as the the *first* Borel–Cantelli lemma while to refer to the nontrivial part (b) as the *second* Borel–Cantelli lemma.

These two results are obtained by Borel (1909, 1912) and Cantelli (1917); see the historical remark at the end of this section. The two Borel–Cantelli lemmas are very useful, indeed often unavoidable, for deducing results about almost sure convergences (the so-called *strong* limit theorems); in particular, these lemmas are needed crucially in establishing the strong laws of large numbers and the laws of iterated logarithms. The first one is more widely applicable since the events there may be completely arbitrary; note that it is valid for any **measure space**—indeed, it holds for an arbitrary **outer measure**, since only the countable subadditivitiy and monotonicity of the set function P were used in the proof of (a) above; see p. 165 of Billingsley (1995), for the relevant definitions.

Remark 1.3.1 The first Borel–Cantelli lemma is a special case of the Monotone Convergence Theorem of measure theory applied to the series of nonnegative terms (see, e.g., Theorem 16.6 of Billingsley (1995)): By the given condition,

$$E\left(\sum_{n=1}^{\infty} I_{A_n}\right) = \sum_{n=1}^{\infty} E(I_{A_n}) = \sum_{n=1}^{\infty} P(A_n) < \infty;$$

so $P\left(\sum_{n=1}^{\infty} I_{A_n} = \infty\right) = 0$. But obviously

$$\left[\sum_{n=1}^{\infty} I_{A_n} = \infty\right] = \limsup A_n. \tag{1.3.3}$$

Indeed, we have shown that if $X_n \geq 0 \; \forall \, n \geq 1$, and $\sum_{n=1}^{\infty} E(X_n) < \infty$, then $\sum_n X_n$ converges with probability one (see Problem 22.3 on p. 294 of Billingsley (1995)).

Remark 1.3.2 The converse of the first Borel–Cantelli lemma would run as follows: If $P(\limsup A_n) = 0$, then $\sum_n P(A_n)$ converges; but this is equivalent to the assertion that if $\sum_{n=1}^{\infty} P(A_n) = \infty$, then $P(\limsup A_n) > 0$. The second Borel–Cantelli lemma shows that, under the **additional** assumption of the **independence of the events** A_n, one has $P(\limsup A_n) = 1$; in this sense, the second Borel–Cantelli lemma is often regarded as a **partial converse** of the first Borel–Cantelli lemma. However, the converse of the first Borel–Cantelli lemma is false as is easily seen from the example: Let $\Omega = (0, 1)$, P be the uniform distribution on Ω, $A_n = (0, 1/n) \; \forall \, n \geq 1$; then $A_n \downarrow \emptyset$ so that $\limsup A_n = \emptyset$, although $\sum P(A_n) = \sum 1/n = \infty$. It should be clear from this example, by taking $A_n = (c, c + 1/n) \forall n \geq m$ where m is a fixed integer $> 1/(1 - c)$, that **if $\sum P(A_n)$ diverges then P(limsupA$_n$) can be any number in [0,1]**; it follows from this observation that the second Borel–Cantelli lemma is false without the assumption of independence (this is also immediate from the example that if $0 < P(A) < 1$ and $A_n = A \forall n \geq 1$, then $\limsup A_n = A$ and $\sum P(A_n) = \infty$; the reader may construct other *nontrivial* examples; see, e.g., Exercise 11 of Chung (2001, p. 82)). A clear picture about the converse of the first Borel–Cantelli lemma can be obtained from the arguments of Remark 1.3.1: Let $N = \sum_{n=1}^{\infty} I_{A_n}$; then $P(N < \infty) = 1$ need not imply $E(N) < \infty$—there are plenty of such random variables N (e.g., define a random variable N by $P(N = n) = cn^{-2} \forall n \geq 1$ where c is a suitable positive real such that $\sum_{n=1}^{\infty} P(N = n) = 1$, and note that $E(N) = \infty$ and, finally, let $A_n = [N \geq n] \forall n \geq 1$ so that $A_n \downarrow$ and $\limsup A_n = \cap_{n=1}^{\infty} A_n = [N = \infty] = \emptyset$, while $\sum_{n=1}^{\infty} P(A_n) = E(N) = \infty$; see Lemma 1.1.1(a)).

Remark 1.3.3 The first Borel–Cantelli lemma is false under the sole (and weaker) assumption that $P(A_n) \to 0$. To see this, let $\Omega = [0, 1]$ and P be the uniform distribution on Ω; let

$$A_n = \left[\frac{n - 2^{k-1}}{2^{k-1}}, \frac{n + 1 - 2^{k-1}}{2^{k-1}} \right] \text{ if } 2^{k-1} \leq n < 2^k \text{ for some } k \geq 1;$$

i.e., $A_1 = [0, 1]$, $A_2 = [0, 1/2]$, $A_3 = [1/2, 1]$, $A_4 = [0, 1/4]$, $A_5 = [1/4, 1/2]$, $A_6 = [1/2, 3/4]$, $A_7 = [3/4, 1]$, and so on. Then it is easy to verify that $P(A_n) \to 0$ and $\limsup A_n = \Omega$. The reader may also verify that *given any $c \in [0, 1]$, there exists a sequence $\{A_n\}_{n \geq 1}$ of events such that $P(A_n) \to 0$ but $P(\limsup A_n) = c$.*

Remark 1.3.4 The second Borel–Cantelli lemma is a consequence of the Kolmogorov strong law of large numbers (SLLN) for a sequence of independent random variables (see, e.g., Theorem II A, p. 250 of Loève (1977)). For, if we set

$$N_n = \sum_{i=1}^{n} I_{A_i}, \quad n \geq 1$$

so that $E(N_n) = \sum_{i=1}^{n} P(A_i) \to \infty$, then, using Lemma 15 on p. 278 of Petrov (1975a), or otherwise directly,

$$\sum_{n=1}^{\infty} \text{var}(I_{A_n})(E(N_n))^{-2} \leq \sum_{n=1}^{\infty} P(A_n) \left(\sum_{i=1}^{n} P(A_i) \right)^{-2} < \infty$$

and hence $(N_n - E(N_n))/E(N_n) \to 0$ a.s., i.e., $N_n/E(N_n) \to 1$ a.s. But then $N_n \to \infty$ with probability 1 since $E(N_n) \to \infty$; i.e., $P(\limsup A_n) = 1$. (See (4.1.2) of Chap. 4, p. 85 for details.)

Historical Remarks

Nash (1954) stated the following remark. See, also, the first paragraph on p. 173 of Móri and Székeley (1983).

Let $P(A_n|I_{A_1}, \ldots, I_{A_{n-1}})$ denote the conditional probability of A_n, given the outcomes of the previous $(n-1)$ trials. When $n = 1$, the expression is $P(A_1)$. The 1912 Borel criterion stated:

If $0 < p'_n \le P(A_n|I_{A_1}, \ldots, I_{A_{n-1}}) \le p''_n < 1$ for every n, whatever be $A_1, \ldots, A_{n-1}$, then $\sum_j p''_j < \infty$ implies that $P(\limsup A_n) = 0$, and $\sum_j p'_j = \infty$ implies that $P(\limsup A_n) = 1$.

Cantelli proved that $\sum_j P(A_j) < \infty$ always implies that $P(\limsup A_n) = 0$. Chung and Erdös (1952) remarked the following.

As Borel already noticed (Borel (1926), p. 48 ff), the assumption of independence in the second Borel–Cantelli lemma can be removed if we assume that

$$\sum_k P(A_k|A_1^c \cap \cdots \cap A_{k-1}^c) = \infty.$$

$\cdots$ Although Borel used this condition successfully in his pioneering work on the metric theory of continued fractions, it is too stringent for many purposes.

The Second Borel–Cantelli Lemma and Subsequences

We conclude this section with one more trivial, but useful, result. In case one wishes to show $P(\limsup A_n) = 1$ where the whole sequence $A_1, A_2, \ldots$ is not independent, it is sometimes possible to get a suitable subsequence of these events which is independent. The following theorem can then be applied; see Examples 1.6.7 and 1.6.8. See Fact (F) on p. 6.

Theorem 1.3.2 *Let $\{A_n\}_{n \ge 1}$ be a sequence of events. Suppose there exists a subsequence of natural numbers, say, $n_1 < n_2 < \cdots$, such that $\{A_{n_k}\}_{k \ge 1}$ are independent and $\sum_k P(A_{n_k}) = \infty$. Then*

$$P(\limsup A_n) = 1.$$

1.4 Some Basic Inequalities

In this section, we collect some probability inequalities which will be used later. Fix a probability space $(\Omega, \mathcal{A}, P)$.

Lemma 1.4.1 (Markov's inequality) *If $X \ge 0$ and $a > 0$, then $P(X \ge a) \le E(X)/a$. A better inequality is*

$$a\, P(X \ge a) \le \int_{[X \ge a]} X\, dP$$

which is valid for any X and any real a. In fact, one has

$$A \in \mathcal{A} \text{ and } A \subset [X \geq a] \Rightarrow aP(A) \leq \int_A XdP,$$

since then a $I_A \leq X\, I_A$.

Example 1.4.1 Let $A_n \in \mathcal{A} \; \forall\, n \geq 1$, and B_m be the event that at least m of these events occur $(m = 1, 2, \ldots)$. Then

$$P(B_m) \leq m^{-1} \sum_{n=1}^{\infty} P(A_n).$$

(If $m = 1$, one gets Boole's inequality).

Solution: Note that $B_m = [N \geq m]$ where $N = \sum_{n=1}^{\infty} I_{A_n}$. By Markov's inequality,

$$P(B_m) = P(N \geq m) \leq m^{-1} E(N) = m^{-1} \sum_{n=1}^{\infty} P(A_n)$$

by the Monotone Convergence Theorem (see, e.g., Theorem 16.6 of Billingsley (1995)). Note that $E(N)$ may be ∞ in this example.

Some **equivalent** forms of Markov's inequality are noted below.

First Alternative Form: *If $a > 0$ and b is real, then for $a > 0$*

$$P(|X - b| \geq a) \leq E(|X - b|^r)/a^r \;\; for\, any\, r > 0.$$

(This is known as the **Chebyshev–Markov inequality**.)

Second Alternative Form: *If $g : [0, \infty) \rightarrow [0, \infty)$ is a function such that $g(a) > 0$ for $a > 0$, and g is nondecreasing, then for $a > 0$*

$$P(|X| \geq a) \leq E(g(|X|))/g(a).$$

The last inequality has an important companion, known as the **Elementary Kolmogorov inequality**, namely,

If $g : [0, \infty) \rightarrow [0, \infty)$ is nondecreasing and $P(g(|X|) \leq M) = 1$ where $M > 0$, then $P(|X| > a) \geq (E(g(|X|)) - g(a))/M$, $a \geq 0$.

For

$$E(g(|X|)) = E(g(|X|) : |X| > a) + E(g(|X|) : |X| \leq a)$$
$$\leq M\, P(|X| > a) + g(a).$$

Here
$$E(g(|X|) : |X| > a) = \int_{[|X|>a]} g(|X|)dP, \text{ etc.}$$

(In general, we shall replace the usual notation $E(XI_A)$ by $E(X : A)$.) As an application of this inequality, we mention the following:

$$E\left(\frac{|X|^r}{1+|X|^r}\right) - \frac{a^r}{1+a^r} \leq P(|X| > a) \leq \frac{1+a^r}{a^r} E\left(\frac{|X|^r}{1+|X|^r}\right) \text{ for } r, a > 0.$$

(Take $g(x) = x^r/(1+x^r)$ for $x \geq 0$.) From this, we get the following well-known result:

$$X_n \to^P X \Leftrightarrow E\left(\frac{|X_n - X|^r}{1+|X_n - X|^r}\right) \to 0 \text{ for some } r > 0 \text{ (or, all } r > 0).$$

For generalizations of Markov's inequality, see Eisenberg and Ghosh (2001).

Chebyshev's inequality

We now turn to the Chebyshev inequality which is, in fact, the special case of the First Alternative Form of Markov's inequality when $r = 2$ and $b = E(X)$ provided $E(|X|) < \infty$. [Often the special case of the First Alternative Form when $r = 2$ is referred to as the **Extended Chebyshev inequality**.]

Lemma 1.4.2 (Chebyshev's inequality) *If $E(|X|) < \infty$, $E(X) = \mu$ and $var(X) = \sigma^2$ then for any $a > 0$*
$$P(|X - \mu| \geq a) \leq \sigma^2/a^2.$$

(*A better inequality is* $a^2 P(|X - \mu| \geq a) \leq E((X - \mu)^2 : |X - \mu| \geq a)$.)

Hölder's and Minkowski's inequalities

Below we put $||X||_p = (E(|X|^p))^{1/p}$ for $p \neq 0$.

We begin with a special case of Jensen's inequality. For this we recall the following definition.

Definition 1.4.1 Let $X : \Omega \to [0, \infty)$ be a random variable on $(\Omega, \mathcal{A}, P)$.

(a) *G* is called the GM (**geometric mean**) of *X* if

$$\log G = E(\log X)$$

provided the right-hand side exists and is finite.

(b) *H* is called the HM (**harmonic mean**) of *X* if

$$\frac{1}{H} = E\left(\frac{1}{X}\right)$$

provided the right-hand side is finite.

Theorem 1.4.1 *Let $X : \Omega \to [a, b]$ be a random variable where $-\infty \le a < b \le \infty$ (If $a = -\infty$, the interval is open at a; similarly, when $b = +\infty$).*

(a) **(A special case of Jensen's inequality)** *If $G : [a, b] \to \mathbb{R}$ is such that $G''(x) \ge 0 \; \forall \; x \in (a, b)$ and $G'(x)$ is continuous on $[a, b]$, and $E(|X|) < \infty$, then $E(G(X))$ exists and*

$$E(G(X)) \ge G(E(X)).$$

(b) *Let $E(|X|) < \infty, a = -\infty$ and $b = +\infty$. Then*

$$E(\exp(X)) \ge \exp(E(X)).$$

In particular,

$$\exp\left(\sum_{i=1}^{\infty} x_i \, p_i\right) \le \sum_{i=1}^{\infty} p_i \exp(x_i) \tag{1.4.1}$$

where $p_i \ge 0 \; \forall \; i \ge 1, \sum_{i=1}^{\infty} p_i = 1$, the x_i are distinct and $\sum_{i=1}^{\infty} |x_i| p_i < \infty$.

Proof (a) By Taylor's theorem

$$G(u) = G(u_0) + (u - u_0)G'(u_0) + \frac{1}{2}(u - u_0)^2 G''(\xi), a \le u \le b, a < u_0 < b,$$

for some $\xi \in (a, b)$. As the third term on the right side is ≥ 0, we must have

$$G(u) \ge G(u_0) + (u - u_0)G'(u_0) \text{ for } a \le u \le b, a < u_0 < b.$$

Now take $u = X(\omega)$ where $\omega \in \Omega$ and $u_0 = E(X)$ which is finite. We consider the case where $a < E(X) < b$; the other cases are easy to deal with. Then

$$G(X(\omega)) \ge G(E(X)) + (X(\omega) - E(X))G'(u_0) \; \forall \; \omega \in \Omega.$$

The expectation of the right-hand side is $G(E(X))$ which is finite. So $E(G(X))$ exists (see Theorem 1.5.9 (b), p. 42, of Ash and Doléans-Dade (2000)), and we have $E(G(X)) \ge G(E(X))$.

(b) follows from (a). □

It is an easy exercise to deduce the following inequality which is valid for $X \ge 0$.

$$H.M. \le G.M. \le A.M. \text{ provided } E(\log X) \text{ is finite.}$$

In (1.4.1), take $x_i = \log a_i$ for $i = 1, \ldots, m$ and $p_i = 0 \; \forall \; i \ge m+1$ and conclude that

$$\prod_{i=1}^{m} a_i^{p_i} \leq \sum_{i=1}^{m} p_i a_i, \text{ provided } 0 < a_i < \infty, \ p_i \geq 0, \ \sum p_i = 1. \qquad (1.4.2)$$

This is the **discrete version** of the so-called **AM-GM inequality**. From this, it follows that

$$a_1 a_2 \ldots a_m \leq \sum_{i=1}^{m} \frac{a_i^{p_i}}{p_i} \text{ if } a_i \geq 0, \ p_i > 1 \text{ and } \sum \frac{1}{p_i} = 1. \qquad (1.4.3)$$

An alternative way to deduce (1.4.3) is to observe that $\exp(x)$ is convex on $\mathbb{R}$ and so

$$a_1 \, a_2 \ldots a_m = \exp\left(\sum_{i=1}^{m} \frac{1}{p_i} \log a_i^{p_i}\right)$$

$$\leq \sum_{i=1}^{m} (1/p_i) \exp(\log a_i^{p_i}) = \sum_{i=1}^{m} a_i^{p_i} / p_i.$$

We next show that

$$ab \geq \frac{a^p}{p} + \frac{b^q}{q} \text{ if } a \geq 0, b \geq 0, \quad p < 1, p \neq 0, \quad q = p/(p-1).$$

To this end, we can assume that $a > 0, b > 0$. By Taylor's theorem,

$$t^m = 1 + m(t-1) + m(m-1)\xi \text{ for } t > 0$$

where ξ is a suitable positive real. If $m > 1$ or $m < 0$, then $m(m-1)\xi > 0$ so that $t^m > mt + (1-m)$ for $t > 0, m > 1$ or $m < 0$. Now take $t = a^p b^{-q}, m = 1/p$ to deduce

$$ab^{-q/p} \geq \frac{1}{p} a^p b^{-q} + \frac{1}{q}.$$

Multiplying both sides by $b^q > 0$ we get the desired inequality.

Theorem 1.4.2 *(a)* (Hölder's inequality) *If $p > 1$ and $\frac{1}{p} + \frac{1}{q} = 1$, then*

$$\|XY\|_1 \leq \|X\|_p \|Y\|_q.$$

The inequality is reversed if $p < 1, p \neq 0$ and $q = p/(p-1)$. [If $p = q = 2$, then Hölder's inequality is known as Cauchy-Schwarz's inequality.]
More generally, if $p_i > 1$ for $i = 1, \ldots, m$ and $\sum \frac{1}{p_i} = 1$, then

$$\left\| \prod_{i=1}^{m} X_i \right\|_1 \leq \prod_{i=1}^{m} \|X_i\|_{p_i}.$$

An equivalent form of the last inequality is

$$E\left(\prod_{i=1}^{m}|X_i|^{\theta_i}\right) \leq \prod_{i=1}^{m}(E(|X_i|))^{\theta_i} \ \ if \ 0 < \theta_i < 1 \ \forall \ i \ and \ \sum \theta_i = 1.$$

*(b) (**Minkowski's inequality**) If $p \geq 1$, then*

$$||X + Y||_p \leq ||X||_p + ||Y||_p.$$

The inequality is reversed if $0 < p < 1$.

For the proofs of (a) and (b) above, see Rudin (1987) and Royden (1988); see, also, Rubel (1964). For some extensions, see Petrov (1995).

We next note that

$$E(|X|^p) \geq (E(|X|))^p \ \ for \ p \geq 1, \tag{1.4.4}$$

the inequality sign being reversed if $p < 1$ and $p \neq 0$. To see this, apply Theorem 1.4.2 (a) with $Y \equiv 1$; an alternative way is to consider the function $G(x) = x^p$ for $x \geq 0$ and use Theorem 1.4.1 (a). The inequality (1.4.4) implies that

$$[E(|X|^p)/(E(|X|))^p]^{1/(1-p)} \leq 1 \ if \ 0 < E(|X|) < \infty \ and \ p > 0, \ p \neq 1, \tag{1.4.5}$$

the inequality being reversed if $p < 0$.

From (1.4.4), one can deduce the **Liapounov inequality**:

$$||X||_r \leq ||X||_s \ if \ 0 < r < s. \tag{1.4.6}$$

To see this, put $p = s/r > 1$, and apply (1.4.4) with $|X|$ replaced by $|X|^r$. For some refinements of (1.4.6), see Petrov (1975b, 2007a,b) and Arnold (1978).

The following example is important in the large deviation theory.

Example 1.4.2 Let $M(t) = E(\exp(tX))$ be the moment generating function of X. Then $M(t)$ is convex and log-convex.

Solution: That $M(t)$ is a convex function follows from the convexity of the exponential function. Next, if $0 < \theta < 1$,

$$M(t\theta + u(1 - \theta))$$
$$= \int \exp(t\theta X + u(1 - \theta)X)dP$$
$$\leq \left(\int \exp(tX)dP\right)^{\theta}\left(\int \exp(uX)dP\right)^{1-\theta} \quad \text{by Hölder's inequality}$$
$$= (M(t))^{\theta}(M(u))^{1-\theta}.$$

Hence $M(t)$ is log-convex.

Theorem 1.4.3 (a) *If $P(X = 0) < 1$ and $E(X)$ is finite, then*

$$P(X \neq 0) \geq (E(|X|))^2/E(X^2),$$

$$P(X \neq 0) \geq ((E(|X|))^p/E(|X|^p))^{1/(p-1)} \ for \, p \neq 1, \quad p \neq 0.$$

(This generalizes the Schwarz inequality, namely, $E(|X|) \leq \sqrt{E(X^2)}$.)
(b) (Paley and Zygmund (1932)) *If $b \leq E(X)$, $E(X)$ is finite and $P(X = 0) < 1$,*

$$P(X > b) \geq (E(X) - b)^2/E(X^2).$$

(c) *Let $S = \sum_{i=1}^n X_i$, $E(X_i)$ be finite for each i and $P(S = 0) < 1$. Then*

$$P(\cup[X_i \neq 0]) \geq \left(\sum_{i=1}^n E(X_i)\right)^2 /E(S^2).$$

(d) (Chung and Erdős (1952)) *If $P(A_1^c \cap \cdots \cap A_n^c) < 1$, then*

$$P\left(\bigcup_{i=1}^n A_i\right) \geq \left(\sum_{i=1}^n P(A_i)\right)/\sum_{i=1}^n \sum_{j=1}^n P(A_i \cap A_j).$$

(e) **(Weighted Chung-Erdős inequality;** Feng et al. (2009)). *Let $A_1, \ldots, A_n$ be events and $w_1, \ldots, w_n$ real weights. If $P\left(\sum_{i=1}^n w_i I_{A_i} = 0\right) < 1$, then*

$$P\left(\bigcup_{i=1}^n A_i\right) \geq \left(\sum_{i=1}^n w_i P(A_i)\right)^2 / \sum_{i=1}^n \sum_{j=1}^n w_i w_j P(A_i \cap A_j).$$

Proof (a) Note that $E(|X|) = E(|X|I_{[X \neq 0]})$. Now apply Hölder's inequality.
(b) We have

$$0 \leq E(X) - b \leq E(X) - E(X : X \leq b) = E(X I_{[X > b]})$$

$$\leq (E(X^2)P(X > b))^{1/2}$$

by Cauchy-Schwarz's inequality.
(c) Note that $[S \neq 0] \subset \cup_{i=1}^n [X_i \neq 0]$.
(d) In (c), take $X_i = I_{A_i} \ \forall \, i$.
(e) In (c), take $X_i = w_i I_{A_i} \ \forall \, i$.   □

For a generalization of the Chung-Erdős inequality, see Petrov (2007b) and Dawson and Sankoff (1967).

Lemma 1.4.1 (Chandra (1999)) *Let c_1, c_2 be non-negative reals, $c_3 \in \mathbb{R}$ and satisfy*

$$P(A_i \cap A_j)$$
$$\leq (c_1 P(A_i) + c_2 P(A_j)) P(A_{j-i}) + c_3 P(A_i) P(A_j) \qquad (1.4.7)$$

whenever $1 \leq i < j \leq n$. Then

$$P\left(\bigcup_{i=1}^{n} A_i\right) \geq \frac{s^2}{s + cs^2 - c_3 \sum_{i=1}^{n}(P(A_i))^2}$$

where $s = \sum_{i=1}^{n} P(A_i) > 0$ and $c = c_3 + 2(c_1 + c_2)$.

Proof As $s > 0$, $P(A_i^c) < 1$ for some i, and so $P(A_1^c \cap \cdots \cap A_n^c) < 1$. By the Chung-Erdös inequality,

$$P\left(\bigcup_{i=1}^{n} A_i\right) \geq s^2 / \sum_{i=1}^{n} \sum_{j=1}^{n} P(A_i \cap A_j) = s^2 / (s + 2 \sum_{1 \leq i < j \leq n} P(A_i \cap A_j)).$$

Next,

$$\sum_{1 \leq i < j \leq n} P(A_i \cap A_j)$$
$$\leq \sum_{1 \leq i < j \leq n} [(c_1 P(A_i) + c_2 P(A_j)) P(A_{j-i}) + c_3 P(A_i) P(A_j)]$$
$$= c_1 \sum_{i=1}^{n-1} P(A_i) \sum_{j=i+1}^{n} P(A_{j-i}) + c_2 \sum_{j=2}^{n} P(A_j) \sum_{i=1}^{j-1} P(A_{j-i})$$
$$+ \frac{1}{2} c_3 \left(s^2 - \sum_{i=1}^{n}(P(A_i))^2\right)$$
$$\leq \left(\frac{1}{2} c_3 + (c_1 + c_2)\right) s^2 - \frac{1}{2} c_3 \sum_{i=1}^{n}(P(A_i))^2. \qquad \square$$

Erdös and Renyi (1959) consider the condition (1.4.7) with $c_1 = c_2 = 0$ and $c_3 = 1$. For other examples, see Kochen and Stone (1964, Examples 1 and 2) and Lamperti (1963, p. 62). See, also, Examples 3.3.1 and 3.3.2.

Remark 1.4.1 If (1.4.7) holds $\forall\, n \geq 1$ and $\sum_{n=1}^{\infty} P(A_n) = \infty$, then $c \geq 1$. For, $s_n^2 \leq \sum_1^n \sum_1^n P(A_i \cap A_j) \leq s_n + cs_n^2 - c_3 \sum_{i=1}^{n}(P(A_i))^2$ (by Schwarz's inequality) where $s_n = \sum_1^n P(A_i)$ now divide by s_n^2 and let $n \to \infty$.

1.5 Applications of the BCL

In this section, we shall discuss some simple and basic applications of the two Borel–Cantelli lemmas. The results discussed below are all well known.

Theorem 1.5.1 (a) *Let* $\epsilon_m \to 0+$ *as* $m \to \infty$. *Then*

$$\sum_{n=1}^{\infty} P(|X_n - X| > \epsilon_m) < \infty \; \forall \, m \geq 1 \Rightarrow X_n \to X \; a.s.;$$

(b) $\sum_{n=1}^{\infty} P(|X_n - X| > \epsilon) < \infty \; \forall \, \epsilon > 0 \Rightarrow X_n \to X \; a.s.;$
(c) $\sum_{n=1}^{\infty} E(|X_n - X|^r) < \infty \; for \; some \; r > 0 \Rightarrow X_n \to X \; a.s.;$
(d) *Let* $\epsilon_n \to 0+$ *as* $n \to \infty$. *Then*

$$\sum_{n=1}^{\infty} P(|X_n - X| > \epsilon_n) < \infty \Rightarrow X_n \to X \; a.s.;$$

$$\sum_{n=1}^{\infty} E(|X_n - X|^r)/\epsilon_n^r < \infty \; for \; some \, r > 0 \Rightarrow X_n \to X \; a.s.;$$

(e) *If* $\sum_{n=1}^{\infty} P(|X_{n+1} - X_n| > \epsilon_n) < \infty$ (*a fortiori, if* $\sum_{n=1}^{\infty} E(|X_{n+1} - X_n|^r)/\epsilon_n^r < \infty$ *for* $r > 0$) *for some sequence* $\{\epsilon_n\}_{n \geq 1}$ *of positive reals such that* $\sum_{n=1}^{\infty} \epsilon_n < \infty$, *then there is a random variable* X *such that* $X_n \to X \; a.s.;$
(f) *Converses of (a) and (d) are false.*

Proof (a) By the first Borel–Canteli lemma,

$$P(|X_n - X| > \epsilon_m \; i.o.(n)) = 0 \; \forall \, m \geq 1.$$

Theorem 1.2.1 now implies that $X_n \to X$ *a.s.*
(b) is a special case of (a).
(c) By the Chebyshev–Markov inequality,

$$P(|X_n - X| > \epsilon) \leq E(|X_n - X|^r)/\epsilon^r \; \forall \, \epsilon > 0.$$

Thus $\sum_{n=1}^{\infty} P(|X_n - X| > \epsilon) < \infty \; \forall \epsilon > 0$ by the given condition. By (b), we then have $X_n \to X$ *a.s.*
Alternative Proof: By the Monotone Convergence Theorem,

$$P\left(\sum_{n=1}^{\infty} |X_n - X|^r < \infty\right) = 1$$

see Remark 1.3.1. But

$$\left[\sum_{n=1}^{\infty} |X_n - X|^r < \infty\right] \subset \left[|X_n - X|^r \to 0\right] = [X_n \to X].$$

So $P(X_n \to X) = 1$.

(d) In view of the first Borel–Cantelli lemma, $P(|X_n - X| > \epsilon_n \ i.o.) = 0$. It is, therefore, enough to show that

$$[X_n \not\to X] \subset \limsup[|X_n - X| > \epsilon_n].$$

To this end, let ω lie outside the set on the right side. Then $\omega \in \liminf[|X_n - X| \le \epsilon_n]$. So $\exists$ an integer $k \ge 1$ such that $|X_n(\omega) - X(\omega)| \le \epsilon_n \ \forall \ n \ge k$; letting $n \to \infty$, we get

$$\limsup_{n\to\infty} |X_n(\omega) - X(\omega)| \le 0, \text{ i.e.}, X_n(\omega) \to X(\omega).$$

(e) Put $A_n = [|X_{n+1} - X_n| > \epsilon_n], n \ge 1$. Then $A_n \in \mathcal{A} \ \forall \ n \ge 1$. By the first Borel–Cantelli lemma, $P(A_n \ i.o.) = 0$, i.e., $P(\liminf A_n^c) = 1$. We argue below that

$$\liminf A_n^c \subset [\lim_n X_n \text{ exists and is finite }].$$

Let $\omega \in \liminf A_n^c$. Then $\exists$ an integer $k \ge 1$ such that

$$|X_{n+1}(\omega) - X_n(\omega)| \le \epsilon_n \ \forall \ n \ge k.$$

So, $\sum_{n=k}^{\infty} |X_n(\omega) - X_{n+1}(\omega)| \le \sum_{n=k}^{\infty} \epsilon_n < \infty$. Hence the series $\sum_{n=1}^{\infty} |X_{n+1}(\omega) - X_n(\omega)|$ converges; i.e.,

$$\lim_{n\to\infty} \sum_{i=1}^{n-1} (X_{i+1}(\omega) - X_i(\omega)) \text{exists and is finite.}$$

This is equivalent to saying that $\lim_n X_n(\omega)$ exists and is finite.

(f) Consider the example: Let X follow the uniform distribution on $(0, 1)$. Put $X_n = 1_{A_n}$ where $A_n = [(n-1)/n < X < 1], n \ge 1$. Then for any $\omega \in (0, 1), \omega \notin A_n$ for all sufficiently large n, and so $X_n(\omega) \to 0$. For any $\epsilon, 0 < \epsilon < 1$, one has $[|X_n| > \epsilon] = [X_n = 1] = A_n$ which implies that

$$\sum_{n=1}^{\infty} P(|X_n| > \epsilon) = \sum_{n=1}^{\infty} P(A_n) = \sum_{n=1}^{\infty} n^{-1} = \infty.$$

Thus the converse of (b) (and hence that of (a)) is false.

Consider again the above example: Let $\epsilon_n \to 0+$. Then $\exists$ an integer $k \geq 1$ such that $\epsilon_n < 1/2 \; \forall \, n \geq k$. Consequently,

$$[|X_n| > \epsilon_n] = [X_n = 1] = A_n \text{ for } n \geq k,$$

and so $\sum_{n=1}^{\infty} P(|X_n| > \epsilon_n) = \infty$. $\qquad\qquad\qquad\qquad\qquad\qquad\quad$ □

Parts (a)–(c) of the above theorem are due to Cantelli.

Theorem 1.5.2 (a) *If* $X_n \to^P X$, *then there is a subsequence* $\{n_k\}_{k \geq 1}$ *of positive integers such that* $X_{n_k} \to X$ *a.s.*
(b) $X_n \to^P X$ *iff given any subsequence* $\{n_k\}_{k \geq 1}$ *of positive integers, there is a further subsequence* $\{n_{k_m}\}$ *of* $\{n_k\}$ *such that* $X_{n_{k_m}} \to X$ *a.s.*

Proof (a) Let $n_1 = 1$, and define $\{n_k\}_{k \geq 1}$ inductively such that

$$P(|X_{n_k} - X| > 2^{-k}) \leq 2^{-k} \; \forall \, k \geq 2.$$

To this end, suppose that $n_1 < n_2 < \cdots < n_{k-1}$ are defined with this property for some $k \geq 2$. As $X_n \to^P X$,

$$P(|X_n - X| > 2^{-k}) \to 0 \text{ as } n \to \infty.$$

So there is an integer $n_k > n_{k-1}$ such that $P(|X_{n_k} - X| > 2^{-k}) \leq 2^{-k}$. Then $\sum_k P(|X_{n_k} - X| > 2^{-k}) < \infty$. By Theorem 1.5.1 (d), $X_{n_k} \to X$ a.s.
(b) 'If' Part:
Suppose, by way of contradiction, that $\{X_n\}_{n \geq 1}$ does not converge X in probability. Then $\exists$ an $\epsilon > 0$ such that $P(|X_n - X| > \epsilon)$ does not tend to 0 as $n \to \infty$. So $\exists \, \delta > 0$ such that

$$P(|X_n - X| > \epsilon) \geq \delta \text{ for infinitely many values of } n.$$

It is now easy to verify that there is a subsequence $\{n(k)\}_{k \geq 1}$ of positive integers such that

$$P(|X_{n(k)} - X| > \epsilon) \geq \delta \; \forall \, k \geq 1. \qquad\qquad (1.5.1)$$

By the given condition, there is a subsequence $\{n(k(m))\}_{m \geq 1}$ of $\{n(k)\}_{k \geq 1}$ such that $X_{n(k(m))} \to X$ a.s. as $m \to \infty$. Hence $X_{n(k(m))} \to^P X$, and so $P(|X_{n(k(m))} - X| > \epsilon) \to 0$ as $m \to \infty$. Thus $\exists$ an integer $m_0 \geq 1$ such that

$$P(|X_{n(k(m))} - X| > \epsilon) < \delta \; \forall \, m \geq m_0. \qquad\qquad (1.5.2)$$

But since $\{n(k(m))\}_{m \geq 1}$ is a subsequence of $\{n(k)\}_{k \geq 1}$, we get from (1.5.1) that

$$P(|X_{n(k(m))} - X| > \epsilon) \geq \delta \; \forall \, m \geq 1. \qquad\qquad (1.5.3)$$

The statements (1.5.2) and (1.5.3) are contradictory.

'Only If' Part:

Let $X_n \to^P X$. Then $X_{n_k} \to^P X$ as well. By (a), there a subsequence $\{n(k(m))\}_{m \geq 1}$ of $\{n_k\}_{k \geq 1}$ and that $X_{n_{k_m}} \to X$ a.s.   □

Theorem 1.5.2 (b) may appear, at a first glance, to be of theoretical value only. However, this is not the case, and this result is very useful. As an illustration, we prove the following theorem. To this end, recall that if $g : \mathbb{R} \to \mathbb{R}$ is any function then the set, D_g, of all discontinuity points of g is an F_σ-set (i.e., a countable union of closed subsets of $\mathbb{R}$), and hence is a Borel set.

Theorem 1.5.3 *If $X_n \to^P X$ and $g : \mathbb{R} \to \mathbb{R}$ is a Borel measureable function such that $P(X \in D_g) = 0$ where D_g is the set of all discontinuity points of g, then $g(X_n) \to^P g(X)$.*

Proof We shall use the full force of Theorem 1.5.2 (b). So let $\{n(k)\}_{k \geq 1}$ be a subsequence of positive integers. By the "sufficiency part" of Theorem 1.5.2 (b), we only need to show the existence of a further subsequence $\{n(k(m))\}_{m \geq 1}$ of $\{n(k)\}_{k \geq 1}$ such that

$$g\left(X_{n(k(m))}\right) \to g(X) \ a.s. \ as \ m \to \infty. \tag{1.5.4}$$

To this end, begin by noting that as $X_n \to^P X$, the "necessary part" of Theorem 1.5.2 (b) implies that there is a subsequence $\{n(k(m))\}_{m \geq 1}$ of $\{n(k)\}_{k \geq 1}$ such that

$$X_{n(k(m))} \to X \ a.s. \ as \ m \to \infty.$$

Let $A = [X_{n(k(m))} \to X]$; then $P(A) = 1$. Let $B = A \cap [X \notin D_g]$.

As $P(X \in D_g) = 0$, $P(B) = 1$. To show (1.5.4), it suffices to show that

$$B \subset [g(X_{n(k(m))}) \to g(X)]. \tag{1.5.5}$$

So let $\omega \in B$. As $\omega \in A$, $X_{n(k(m))}(\omega) \to X(\omega)$. This fact and the fact that $\omega \in [X \notin D_g]$, i.e., that g is continuous at $X(\omega)$ imply that ω lies in set on the right side of (1.5.5).   □

Example 1.5.1 Let $\{X_n\}_{n \geq 1}$ be any sequence of random variables defined on the same probability space. Show that $\exists$ a sequence $\{a_n\}_{n \geq 1}$ of positive reals such that $X_n/a_n \to 0 \ a.s$.

Solution: Fix an integer $n \geq 1$. Clearly, $\exists d_n > 0$ such that $P(|X_n| > d_n) \leq n^{-2}$. Put $a_n = nd_n > 0, n \geq 1$. Then $\sum_{n=1}^{\infty} P(|X_n|/a_n > 1/n) < \infty$. By Theorem 1.5.1 (d), $X_n/a_n \to 0$ a.s. [If the X_n are defined on different probability spaces, $\exists \{a_n\}_{n \geq 1}$ such that $a_n > 0 \ \forall \ n \geq 1$ and $X_n/a_n \to^P 0$.]

Example 1.5.2 Let $\sum P(X_n > ca_n) < \infty$ where $c > 0$, $a_n > 0$. Then show that

$$\limsup \frac{X_n}{a_n} \leq c \ a.s.$$

Solution: By the first Borel–Cantelli lemma,

$$P(X_n > c \, a_n \ i.o.) = 0, \text{ i.e., } P(X_n \leq c \, a_n \text{ eventually}) = 1.$$

But $[X_n \leq c \, a_n \text{ eventually }] \subset [\limsup \frac{X_n}{a_n} \leq c]$.

In the next section, we shall give some typical examples on the Borel–Cantelli lemmas. For the application of the second one, we shall make a remark. First a definition is needed.

Recall that a sequence $\{X_n\}_{n\geq 1}$ is called **pairwise negative quadrant dependent (pairwise NQD)** if $\forall i \neq j, s, t \in \mathbb{R}$

$$P(X_i > s, X_j > t) \leq P(X_i > s)P(X_j > t).$$

A sequence $\{A_n\}_{n\geq 1}$ of events is called **pairwise NQD** if $\{I_{A_n}\}_{n\geq 1}$ is so. Clearly, $\{A_n\}_{n\geq 1}$ is pairwise NQD iff

$$P(A_i \cap A_j) \leq P(A_i)P(A_j) \ \forall i \neq j. \tag{1.5.6}$$

A sequence $\{A_n\}_{n\geq 1}$ of pairwise independent events is pairwise NQD ((1.5.6) then holds with equality in place of $\leq$). Also, if $\{A_n\}_{n\geq 1}$ are pairwise NQD, so are $\{A_n^c\}_{n\geq 1}$.

Remark 1.5.1 The second Borel–Cantelli lemma holds if 'independence of $\{A_n\}$' is replaced by '$\{A_n\}$ is pairwise NQD'. This was first noted by Erdös and Renyi (1959), and can be proved in several ways. These will be discussed in Chap. 3. Variants of this result will be proved in Chap. 3 where other dependence conditions on $\{A_n\}_{n\geq 1}$ will be assumed and one can then only conclude that $P(\limsup A_n) > 0$, and this suffices in many examples. These facts should be remembered carefully while going through the examples of the next section.

Theorem 1.5.4 *Let $\{X_n\}_{n\geq 1}$ be pairwise independent with the distribution functions $\{F_n\}_{n\geq 1}$. Then $X_n \to 0$ a.s. $\Leftrightarrow \sum P(|X_n| > \epsilon) < \infty \ \forall \epsilon > 0 \Leftrightarrow \sum(1 - F_n(\epsilon) + F_n(-\epsilon)) < \infty \ \forall \epsilon > 0$. Also $\limsup |X_n| = \infty$ a.s. $\Leftrightarrow \sum P(|X_n| > \epsilon) = \infty \ \forall \epsilon > 0$.*

Proof Note that

$$X_n \to 0 \ a.s. \Leftrightarrow P(|X_n| > \epsilon \ i.o.) = 0 \ \forall \epsilon > 0$$

$$\Leftrightarrow \sum P(|X_n| > \epsilon) < \infty \ \forall \epsilon > 0$$

by the Borel–Cantelli lemmas.

If $\sum(1 - F_n(\epsilon) + F_n(-\epsilon)) < \infty \forall \epsilon > 0$, then $\sum P(|X_n| > \epsilon) < \infty \forall \epsilon > 0$ since $P(|X_n| > \epsilon) \le 1 - F_n(\epsilon) + F_n(-\epsilon)$, and so $X_n \to 0$ a.s If $X_n \to 0$ a.s., then $P(|X_n| \ge \epsilon \, i.o.) = 0 \forall \epsilon > 0$ and so $\sum P(|X_n| \ge \epsilon) < \infty \forall \epsilon > 0$ implying $\sum(1 - F_n(\epsilon) + F_n(-\epsilon)) < \infty \forall \epsilon > 0$.

Finally, $\sum P(|X_n| > \epsilon) = \infty \ \forall \ \epsilon > 0 \Leftrightarrow P(|X_n| > \epsilon \, i.o.) = 1 \forall \epsilon > 0$. $\Leftrightarrow P\left(\bigcap_{m=1}^{\infty}[|X_n| > m \, i.o.(n)]\right) = 1 \Leftrightarrow P(\limsup |X_n| = \infty) = 1$.

For some additional applications of the Borel–Cantelli lemmas, see pp. 100–104 of Gut (2005).

1.6 Examples

Varieties of problems can be solved by the Borel–Cantelli lemmas. We discuss below a few of them.

Example 1.6.1 Assume that each X_n takes only finitely many (*distinct*) values, say, $a, x_1, \ldots, x_m$. Then

(a) $X_n \to^P a \Leftrightarrow P(X_n \ne a) \to 0$;
(b) $X_n \to a$ a.s., if $\sum P(X_n \ne a) < \infty$;
(c) Converse of (b) holds if the events $[X_n = a]$ are pairwise NQD; more precisely, $\sum P(X_n \ne a) = \infty \Rightarrow P(X_n \to a) = 0$ (under this assumption);
(d) if $\{A_n\}$ is pairwise NQD, $I_{A_n} \to 0$ a.s. $\Leftrightarrow \sum P(A_n) < \infty$;
(e) let $m = 1, a < x_1$ and show that $P(\liminf X_n = a, \limsup X_n = x_1) = 1 \Rightarrow \sum P(X_n = a) = \infty$ and $\sum P(X_n = x_1) = \infty$; the reverse implication is true if the events $[X_n = a]$ are pairwise NQD. (In case X_n's are Bernoulli variables and the X_n are pairwise independent with $P(X_n = 1) \to 0$ and $\sum P(X_n = 1) = \infty$, we may then conclude that $\{X_n\}_{n \ge 1}$ continues to visit both 0 and 1 over and over infinitely often with probability one).

Solution:

(a) Let $\epsilon = \frac{1}{2}\min\{|a - x_1|, \ldots, |a - x_m|\} > 0$. Then

$$P(|X_n - a| > \epsilon) \ge P(X_n \ne a).$$

This proves the implication $\Rightarrow$; the reverse implication is always true.
(b) This is always true. By the first Borel–Cantelli lemma, $P(X_n = a \text{ eventually})=1$. So $P(X_n \to a) = 1$. Alternatively, $\infty > \sum_1^\infty P(A_n) = E\left(\sum I_{A_n}\right)$ and so $\sum I_{A_n} < \infty$ a.s. implying that $I_{A_n} \to 0$ a.s.; but then $X_n \to a$ a.s. Here $A_n = [X_n \ne a]$.
(c) By the second Borel–Cantelli Lemma in conjunction with Remark 1.5.1,

$$\sum P(X_n \ne a) = \infty \Rightarrow P(X_n \ne a \, i.o.) = 1 \Leftrightarrow P(X_n = a \text{ eventually}) = 0.$$

Now note that $[X_n \to a] = [X_n = a \text{ eventually}]$.

(d) See the solution of (c).

(e) Assume the given conditions. Then

$$P([X_n = a \, i.o.] \cap [X_n = x_1 \, i.o.]) = 1.$$

The implication $\Rightarrow$ follows by the first Borel–Cantelli lemma. The reverse implication is true by the second Borel–Cantelli lemma.

Example 1.6.2 $X_n \to^P 0 \not\Rightarrow X_n \to 0 \, a.s.$

Solution: Let $P(X_n = 0) = \frac{1}{n} = 1 - P(X_n = 1)$ for $n \geq 1$, and let $\{X_n\}_{n \geq 1}$ be pairwise independent. Then $X_n \to^P 0$ by Example 1.6.1(a), and lim sup $X_n = 1$ *a.s.* by the solution of Example 1.6.1.(c).

Example 1.6.3 (Shuster (1970)) The following are equivalent:

(a) For each $\epsilon > 0$, $\exists$ an event A such that $P(A) \geq 1 - \epsilon$ and $\sum_{n=1}^{\infty} P(A_n \cap A) < \infty$.

(b) $P(\limsup A_n) = 0$.

(c) $P(\liminf A_n^c) = 1$.

Solution: $(a) \Rightarrow (b)$ Let $\epsilon > 0$. If suffices to show that $P(\limsup A_n) \leq \epsilon$. By (a), $\exists$ an event A such that $P(A) \geq 1 - \epsilon$ and $\sum_{n=1}^{\infty} P(A_n \cap A) < \infty$. Then $P(\limsup(A_n \cap A)) = 0$. But then

$$P(\limsup A_n) \leq P((\limsup A_n) \cap A) + P(A^c) \leq \epsilon.$$

(b) $\Rightarrow$ (c): Obvious. (c) $\Rightarrow$ (a): Let $\epsilon > 0$. By (c), $P\left(\cap_{k=n}^{\infty} A_k^c\right) \uparrow 1$. So $\exists$ an integer $n \geq 2$ such that $P\left(\cap_{k=n}^{\infty} A_k^c\right) \geq 1 - \epsilon$. Put $A = \cap_{k=n}^{\infty} A_k^c$. Then $P(A) \geq 1 - \epsilon$, and

$$\sum_{m=1}^{\infty} P(A_m \cap A) = \sum_{m=1}^{n-1} P(A_m \cap A) < \infty.$$

[We have, therefore, shown that if $\sum_n P(A \cap A_n) < \infty$, then $P(\limsup A_n) \leq 1 - P(A)$. It then follows that

$$P(\limsup A_n) = 1 - \sup\{P(A) : A \in \mathcal{F}\}$$

where $\mathcal{F}$ is the set of all events A satisfying $\sum_{n=1}^{\infty} P(A \cap A_n) < \infty$; for, the sets $\cap_{k=n}^{\infty} A_k^c$ for each $n \geq 1$ belong to $\mathcal{F}$, and

$$P(\liminf A_n^c) = \sup_{n \geq 1} P\left(\bigcap_{k=n}^{\infty} A_k^c\right).]$$

Example 1.6.4 Let $\{X_n\}$ be pairwise independent. Show that for each real a,

$$P(X_n \to a) = 0 \text{ or } 1.$$

Solution: Fix a real a. Note that by (1.2.16)

$$[X_n \to a] = \bigcap_{m=1}^{\infty} \liminf_{n \to \infty} [|X_n - a| \le 1/m] = \bigcap_{m=1}^{\infty} \liminf_{n \to \infty} A_{n,m} \text{ (say) .} \quad (1.6.1)$$

Case 1 For each $m \ge 1$, we have $\sum_{n=1}^{\infty} P(A_{n,m}^c) < \infty$. Then $P(A_{n,m}^c \text{ i.o. } (n)) = 0 \, \forall \, m \ge 1$ which implies that $P(\liminf_{n \to \infty} A_{n,m}) = 1 \, \forall \, m \ge 1$, and so $P(X \to a) = 1$ by (1.6.1).

Case 2 $\exists$ an integer $m \ge 1$ such that $\sum_{n=1}^{\infty} P(A_{n,m}^c) = \infty$. Clearly, $\{A_{n,m}^c\}_{n \ge 1}$ are pairwise independent. So

$$P(A_{n,m}^c \text{ i.o.}(n)) = 1, \text{ i.e. }, P(\liminf_{n \to \infty} A_{n,m}) = 0.$$

By (1.6.1), $[X_n \to a] \subset \liminf_{n \to \infty} A_{n,m}$; so $P(X_n \to a) = 0$.

Example 1.6.5 Let P, Q be two probabilities on a measurable space $(\Omega, \mathcal{A})$. Then the following are equivalent:

(a) $A \in \mathcal{A}$ and $P(A) = 0 \Rightarrow Q(A) = 0$.
(b) For each $\epsilon > 0$, $\exists \delta > 0$ such that $A \in \mathcal{A}$ and $P(A) < \delta \Rightarrow Q(A) < \epsilon$.
(c) $A_n \in \mathcal{A} \, \forall \, n \ge 1$ and $P(A_n) \to 0 \Rightarrow Q(A_n) \to 0$.

Solution:

(a) $\Rightarrow$ (b) Suppose that the negation of (b) holds. Then $\exists \epsilon > 0$ such that for each $\delta > 0$ there exists an event $A \in \mathcal{A}$ with the property that $P(A) < \delta$ but $Q(A) \ge \epsilon$. Taking $\delta = 1, (\frac{1}{2})^2, (\frac{1}{3})^2, \ldots$ successively, we get a sequence $\{A_k\}_{k \ge 1}$ of events in $\mathcal{A}$ such that $P(A_k) < k^{-2}$ and $Q(A_k) \ge \epsilon \, \forall \, k \ge 1$. So $P(\limsup A_k) = 0$ and $Q(\limsup A_k) \ge \limsup Q(A_k) \ge \epsilon > 0$. Thus the negation of (a) holds.

(b) $\Rightarrow$ (c): Let $A_n \in \mathcal{A} \, \forall \, n \ge 1$ and $P(A_n) \to 0$. We show that $Q(A_n) \to 0$. Let $\epsilon > 0$. In view of (b), $\exists \delta > 0$ such that $A \in \mathcal{A}$ and $P(A) < \delta \Rightarrow Q(A) < \epsilon$. As $P(A_n) \to 0$, $\exists$ an integer $m \ge 1$ such that $P(A_n) < \delta \, \forall \, n \ge m$. So $Q(A_n) < \epsilon \, \forall \, n \ge m$. Hence $Q(A_n) \to 0$.

(c) $\Rightarrow$ (a): Let $A \in \mathcal{A}$ and $P(A) = 0$. Put $A_n = A \, \forall \, n \ge 1$. Then $P(A_n) \to 0$. So by (c), $Q(A_n) \to 0$; i.e., $Q(A) = 0$.

Example 1.6.6 Let $\{X_n\}_{n \ge 1}$ be pairwise independent and identically distributed. Let $E(|X_1|) = \infty$. Show that if $S_n = X_1 + \cdots + X_n$ and $\bar{X}_n = n^{-1} S_n, n \ge 1$,

(a) For each $a > 0$, $P(\limsup[|X_n| > na]) = 1$;
(b) For each $a > 0$, $P(\limsup[|S_n| > na]) = 1$; and

(c) $P(\lim \bar{X}_n$ exists and is finite$) = 0$.
     [Indeed, $P(\limsup |\bar{X}_n| = \infty) = 1$.]

Solution:

(a) Fix a real a. Since the events $\{[|X_n| > na]\}_{n\geq 1}$ are pairwise independent, it suffices to show that $\sum_{n=1}^{\infty} P(|X_n| > na) = \infty$. As the X_n are identically distributed, it suffices to show that $\sum_{n=1}^{\infty} P(|X_1| > na) = \infty$. But

$$\sum_{n=1}^{\infty} P(|X_1| > na) \geq \sum_{n=1}^{\infty} P(|X_1/a| \geq n+1) \geq E(|X_1/a|) - 1 - P(|X_1| \geq a) = \infty.$$

(b) Fix a real a. Note that

$$[|S_n| \leq na \text{ eventually}] \subset [|X_n| \leq 2na \text{ eventually}].$$

For, if ω lies in the set on the left side, then $\exists$ an integer $m \geq 1$ such that $|S_n(\omega)| \leq na \,\forall n \geq m$, and then

$$|S_{n-1}(\omega)| \leq (n-1)a \leq na \quad \forall n \geq +1$$

so that

$$|X_n(\omega)| = |S_n(\omega) - S_{n-1}(\omega)| \leq 2na \quad \forall n \geq +1.$$

Consequently,

$$P(\limsup[|S_n| > na]) \geq P(\limsup[|X_n| > 2na]) = 1 \text{ by (a)}.$$

(c) It suffices to note that

$$[\,\lim \bar{X}_n \text{ exists and is finite}\,] \subset [\,\limsup |\bar{X}_n| \text{ is finite}\,].$$

[Under the assumptions of Example 1.6.6, it is possible that $\{n^{-1}S_n\}$ is stochastically bounded; consider, e.g., the Cauchy distribution.]

Example 1.6.7 Let $P(X_n = 1) = p = 1 - P(X_n = 0)$ with $0 < p < 1$. Let the X_n be independent. Let β be a fixed $k \times 1$ vector whose components are 0 or 1. Show that $P((X_n, \ldots, X_{n+k-1}) = \beta \text{ i.o. } (n)) = 1$.

Solution: Let $A_n = [(X_n, \ldots, X_{n+k-1}) = \beta]$, $B_n = [(X_{(n-1)k+1}, \ldots, X_{nk}) = \beta]$, $n \geq 1$. Obviously, the events $\{B_n\}_{n\geq 1}$ are independent (but the events $\{A_n\}_{n\geq 1}$ are **not**). Also, $[B_n \text{ i.o.}] \subset [A_n \text{ i.o.}]$. It, therefore, suffices to show that $\sum P(B_n) = \infty$. As $P(B_n) = P(B_1) > 0 \,\forall n \geq 1$, we are done. (See Theorem 1.3.2.)

   [At this stage, it is instructive to read the story on **"The Monkey and the Type-writter"** in Sect. 18.2 of Chap. 2 of Gut (2005).]

Example 1.6.8 (Breiman (1968)) Consider Example 1.6.7. Let $Y_n = 2X_n - 1, n \geq 1$ (i.e., in terms of successive independent coin-tossing experiment, $Y_n = +1$ or -1 according as the nth toss leads to a head or a tail). Put $S_n = Y_1 + \cdots + Y_n, n \geq 1$. Show that

(a) if $p \neq 1/2$, then $P(S_n = 0 \ i.o.) = 0$; and
(b) if $p = 1/2$, then $P(S_n = 0 \ i.o.) = 1$.

Solution: (a) Let $\bar{S}_n = X_1 + \cdots + X_n, n \geq 1$. Below we write $a_n \sim b_n$ in case $a_n/b_n \to 1$. Then

$$P(S_{2n} = 0) = P(\bar{S}_{2n} = n)$$
$$= \binom{2n}{n} p^n (1 - p)^n$$
$$\sim \frac{(2n)^{2n+1/2} \exp(-2n)}{\sqrt{2\pi} \left(n^{n+1/2} \exp(-n)\right)^2} (p(1-p))^n$$
$$= (n\pi)^{1/2} (4p(1-p))^n,$$

where we have used Stirling's formula for factorials, namely,

$$n! \sim \sqrt{2\pi} n^{n+1/2} \exp(-n).$$

Thus $\sum P(S_{2n} = 0) \leq \sum \pi^{1/2} (4p(1-p))^n < \infty$ as $0 < p < 1$. Thus $P(S_{2n} = 0 \ i.o.) = 0$. Since

$$S_n = 0 \Rightarrow n \text{ is even },$$

$$[S_{2n} = 0 \ i.o.] = [S_n = 0 \ i.o.]. \text{ So } P(S_n = 0 \ i.o.) = 0.$$

(b) We first show that $\exists$ a sequence $\{r(n)\}_{n \geq 1}$ of integers ≥ 1 such that $P(|S_{r(n)}| < n) \leq 1/2 \ \forall \ n \geq 1$.
To this end, let i be an integer and put $q_n = P(S_n = i)$. Then

$$q_n = P(\bar{S}_n = (n+i)/2)$$
$$= \frac{n!}{((n+i)/2)!((n-i)/2)!} (1/2)^n, \text{ provided } (n+i) \text{ is even.}$$

Hence for large n with $(n+i)$ even, we can write by Stirling's formula

$$\log q_n = -\log(2\pi)/2 + (n + 1/2) \log n - ((n+i+1)/2) \log((n+i)/2)$$
$$- ((n-i+1)/2) \log((n-i)/2) - n \log 2 + o(1)$$
$$= -\log(2\pi)/2 + \log 2 - (\log n)/2 - ((n+i+1)(i/(2n) + o(n^{-1}))$$

$$- (n - i + 1)(-i/(2n) + o(n^{-1})) + o(1)$$
$$= - \log(2\pi)/2 + \log 2 - (\log n)/2 + o(1) \to -\infty.$$

Hence $q_n \to 0 \, \forall i \geq 1$. Thus

$$P(|S_n| < k) = \sum_{i:|i|<k} P(|S_n| = i) \to 0 \quad \forall k \geq 1.$$

Thus given any $k \geq 1, \exists$ an integer $r(k) \geq 1$ such that $P(|S_{r(k)}| < k) \leq 1/2$.

Now let $n_1 = 1, \ m_k = n_k + r(n_k)$ and $n_{k+1} = m_k + r(m_k)$ for $k \geq 1$. Then $n_k < m_k < n_{k+1} \ \forall k \geq 1$. Define

$$B_k = \left[\sum_{i=n_k+1}^{m_k} Y_i \leq -n_k, \ \sum_{i=m_k+1}^{n_{k+1}} Y_i \geq m_k \right], k \geq 1.$$

Then $B_k \subset [S_n = 0$ for some n with $m_k \leq n \leq n_{k+1}]$. For, if $\omega \in B_k$, $S_{m_k}(\omega) = \sum_{i=1}^{n_k} Y_i + \sum_{i=n_k+1}^{m_k} Y_i \leq n_k + (-n_k) = 0$, as well as, $S_{n_{k+1}}(\omega) \geq -m_k + m_k = 0$. Thus $[B_n \ i.o.] \subset [S_n = 0 \ i.o.]$.

It is, therefore, enough to show that $P(B_n \ i.o.) = 1$. Clearly, the events $\{B_k\}_{k \geq 1}$ are independent (and the events $\{[S_n = 0]\}_{n \geq 1}$ are **not**). The proof will be complete if we show that $\sum P(B_n) = \infty$. But

$$P(B_n) = P \left(\sum_{i=n_k+1}^{m_k} Y_i \leq -n_k \right) P \left(\sum_{i=m_k+1}^{n_{k+1}} Y_i \geq m_k \right)$$

$$= \frac{1}{4} P \left(| \sum_{i=n_k+1}^{m_k} Y_i | \geq n_k \right) P \left(| \sum_{i=m_k+1}^{n_{k+1}} Y_i | \geq m_k \right)$$

$$= \frac{1}{4} P \left(|S_{m_k - n_k}| \geq n_k \right) P \left(|S_{n_{k+1} - m_k}| \geq m_k \right)$$

$$= \frac{1}{4} P \left(|S_{r(n_k)}| \geq n_k \right) P \left(|S_{r(m_k)}| \geq m_k \right)$$

$$\geq 1/16 \text{ by the definition of } \{r(n)\}_{n \geq 1}.$$

Example 1.6.9 (a) Let $P(|X_n| > a) \leq P(|Y| > a)$ for each $a > 0$ and $n \geq 1$ (*a fortiori*, let $\{X_n\}$ be identically distributed). Put $Y_n = n^{-1}X_n$. Show that $Y_n \to^P 0$.

(b) If $\{X_n\}$ are pairwise independent and identically distributed and $Y_n = n^{-1}X_n$, then $Y_n \to 0$ a.s. $\Leftrightarrow E(|X_1|) < \infty$.

Solution: (a) Let $\epsilon > 0$. Note that

$$P(|Y_n| > \epsilon) = P(|X_n| > n\epsilon) \leq P(|Y| > n\epsilon) \to 0.$$

(b) We shall use Lemma 1.1.1(b), p. 4. Now

$$Y_n \to 0 \, a.s. \Leftrightarrow P(|Y_n| \geq \epsilon \, i.o.) = 0 \, \forall \epsilon > 0 \text{ by Theorem 1.2.1}$$

$$\Leftrightarrow \sum_{n=1}^{\infty} P(|Y_n| \geq \epsilon) < \infty \, \forall \epsilon > 0$$

by the Borel–Cantelli lemmas

$$\Leftrightarrow \sum_{n=1}^{\infty} P(|X_n| \geq n\epsilon) < \infty \, \forall \epsilon > 0$$

$$\Leftrightarrow \sum_{n=1}^{\infty} P(|X_1| \geq n\epsilon) < \infty \, \forall \epsilon > 0$$

$$\Leftrightarrow E(|X_1/\epsilon|) < \infty \, \forall \epsilon > 0$$

$$\Leftrightarrow E(|X_1|) < \infty.$$

Example 1.6.10 Let $\{X_n\}_{n \geq 1}$ be iid and put $Y_n = n^{-1} \max_{1 \leq i \leq n} |X_i|$. Then

(a) $Y_n \to^P 0 \Leftrightarrow nP(|X_1| > n) \to 0$;

(b) $Y_n \to 0 \, a.s. \Leftrightarrow E(|X_1|) < \infty$; and

(c) if $E(|X_1|) < \infty$ and $E(X_1) \neq 0$, $\max_{1 \leq i \leq n} |X_i| / \left(\sum_{i=1}^{n} X_i \right) \to 0 \, a.s.$.

Solution:

(a) We first prove the following preliminary results.

(i) $n \, P(|X_1| > n) \to 0 \Rightarrow x P(|X_1| > x) \to 0$ as $x \to \infty$.

(ii) If $0 \leq a_n \leq 1 \, \forall n \geq 1$, then $na_n \to 0 \Leftrightarrow (1 - a_n)^n \to 1$.

To show (i), note that if $x \geq 1$

$$x P(|X_1| > x) \leq 2[x]P(|X_1| > x) \leq 2[x]P(|X_1| > [x]) \to 0 \text{ as } x \to \infty.$$

To show (ii), note that if $na_n \to 0$

$$|1 - (1 - a_n)^n| \leq na_n \to 0$$

as $(1 - h)^n \geq 1 - nh$ for $h \geq 0$,

while if $(1 - a_n)^n \to 1$, then $n \log(1 - a_n) \to 0$, and so $na_n \leq -n \log(1 - a_n) \to 0$. (Recall that $\log x \leq x - 1$ for $0 \leq x < \infty$.)

We now prove (a). If $Y_n \to^P 0$, then

$$P(|Y_n| > 1) \to 0 \Rightarrow P(|Y_n| \leq 1) \to 1$$
$$\Rightarrow (P(|X_1| \leq n))^n \to 1$$

$$\Rightarrow (1 - P(|X_1| > n))^n \rightarrow 1$$
$$\Rightarrow n \, P(|X_1| > n) \rightarrow 0 \text{ by (ii) above.}$$

Conversely, if $nP(|X_1| > n) \rightarrow 0$ and $\epsilon > 0$

$$P(|Y_n| > \epsilon) = 1 - (1 - P(|X_1| > n\,\epsilon))^n \rightarrow 1 - 1 = 0 \text{ (by (ii) above)}$$

since $x \, P(|X_1| > x) \rightarrow 0$ as $x \rightarrow \infty$ (by (i) above) and so $n \, P(|X_1| > n\,\epsilon) \rightarrow 0$.

(b) First note that if $0 < s_n \uparrow \infty$, then

$$a_n/s_n \rightarrow 0 \Rightarrow \max_{1 \le i \le n} |a_i|/s_n \rightarrow 0; \tag{1.6.2}$$

for, if $\epsilon > 0$, $\exists$ an integer $m \ge 1$ such that $|a_n|/s_n < \epsilon \; \forall n \ge m$, which implies that for $n \ge m$

$$\left(\max_{1 \le i \le n} |a_i| \right) / s_n = \left(\max_{1 \le i \le m} |a_i| \right) / s_n + \left(\max_{m \le i \le n} |a_i| \right) / s_n$$
$$< \left(\max_{1 \le i \le m} |a_i| \right) / s_n + \epsilon$$

which, in turn, implies that

$$\limsup \left(\left(\max_{1 \le i \le n} |a_i| \right) / s_n \right) \le \epsilon$$

and so we are done.

We now establish (b). Note that

$$E(|X_1|) < \infty \Leftrightarrow n^{-1} X_n \rightarrow 0 \, a.s. \text{ by Example 1.6.9 (b)}$$
$$\Leftrightarrow \left(\max_{1 \le i \le n} |X_i| \right) / n \rightarrow 0 \, a.s. \text{ by (1.6.2).}$$

(c) The SLLN as given in p. 282 of Billingsley (1995) will be usual. By (b),

$$n^{-1} \max_{1 \le i \le n} |X_i| \rightarrow 0 \, a.s.$$

By the above SLLN, $n^{-1} S_n \rightarrow E(X_1) \, a.s.$ As $E(X_1) \ne 0$, we are done.

Remark 1.6.1 Let $\{X_n\}_{n \ge 1}$ be pairwise independent and identically distributed and $\alpha > 0$. Then $n^{-1/\alpha} X_n \rightarrow 0 \, a.s.$ iff $E(|X_1|^\alpha) < \infty$, and $n^{-1/\alpha} X_n \rightarrow \infty \, a.s.$ iff $E(|X_1|^\alpha) = \infty$. [This is a sort of a **zero-one** law.]

Remark 1.6.2 For a sequence, $Y, X_1, X_2, \ldots$, of random variables and a real $r > 0$, consider the following statements:

(a) $n^{-1/r} X_n \to 0 \, a.s.$
(b) $P(|X_n| > n^{1/r} \epsilon \, i.o.) = 0 \quad \forall \epsilon > 0.$
(c) $\sum P(|Y| > n^{1/r} \epsilon) < \infty \quad \forall \epsilon > 0.$
(d) $\sum P(|X_1| > n^{1/r} \epsilon) < \infty \quad \forall \epsilon > 0.$
(e) $E(|Y|^r) < \infty.$
(f) $E(|X_1|^r) < \infty.$
(g) $n^{-1/r} \max_{1 \le i \le n} |X_i| \to 0 \, a.s.$

Then $(g) \Leftrightarrow (a) \Leftrightarrow (b)$, and $(e) \Leftrightarrow (c)$, $(f) \Leftrightarrow (d)$ hold. If

$$P(|X_n| > x) \le P(|Y| > x) \quad \forall \, x > 0, n \ge 1,$$

then $(c) \Rightarrow (b)$ holds. If $\{X_n\}_{n \ge 1}$ are pairwise independent and identically distributed, then $(b) \Leftrightarrow (d)$ holds.

[See the solution of Example 1.6.10(b) and Remark 1.6.1.]

Example 1.6.11 If $\{X_n\}_{n \ge 1}$ are pairwise independent and

$$a_n^{-1}(X_1 + \cdots + X_n) \to 0 \, a.s.$$

where $a_n > 0$ and $\{a_{n-1}/a_n\}_{n \ge 1}$ is bounded, then $\sum P(|X_n| \ge a_n) < \infty.$

Solution: Clearly,

$$a_n^{-1} X_n = a_n^{-1}(X_1 + \cdots + X_n) - (a_{n-1}/a_n) a_{n-1}^{-1}(X_1 + \cdots + X_{n-1}) \to 0 \, a.s.$$

Put $A = [a_n^{-1} X_n \to 0]$. Then $P(A) = 1$, and $A \subset [a_n^{-1}|X_n| < 1 \text{ eventually}]$. Thus $P(|X_n| \ge a_n \, i.o.) = 0.$

Suppose the conclusion is false. Then $\sum P(|X_n| \ge a_n) = \infty$ and so $P(|X_n| \ge a_n \, i.o.) = 1$, which is a contradiction. [The conclusion holds if $a_n^{-1}(X_1 + \cdots + X_n) \to \mu \, a.s.$ and $a_{n-1}/a_n \to 1$.]

Example 1.6.12 Show that there are examples of independent random variables $\{Y_n\}_{n \ge 1}$ such that $Y_n \to^P 0$ and $P(\limsup Y_n = \infty, \liminf Y_n = -\infty) = 1.$

Solution: Let the $\{X_n\}_{n \ge 1}$ be independent and X_n follow Bin$(1; 1/n), n \ge 1$. Put $Y_n = (-1)^n n \, X_n$. Then $P(Y_n \ne 0) = 1/n \to 0$ but $\sum P(Y_{2n} = 2n) = \infty$ and $\sum P(Y_{2n+1} = -(2n+1)) = \infty$. So the second Borel–Cantelli lemma implies that $P(\limsup Y_n = \infty, \liminf Y_n = -\infty) = 1$, [It is now clear how to construct $\{Y_n\}_{n \ge 1}$ with these properties and having absolutely continuous distributions.]

Example 1.6.13 (Chow and Teicher (1997))

(a) If $\{A_n\}_{n \ge 1}$ and $\{B_n\}_{n \ge 1}$ are two sequences of events satisfying $\sum P(A_n) = \infty$ and $\exists$ an integer $k \ge 1$ such that

$$P\left(A_i \cap \left(\bigcup_{j=1}^{\infty} A_{i+jk}\right)\right) \leq P(A_i)P\left(\bigcup_{j=k}^{\infty} B_j\right) \quad \text{for all sufficiently large} i,$$

then $P\left(\bigcup_{j=k}^{\infty} B_j\right) = 1$.

(b) Let $\{X_n\}_{n\geq 1}$ be iid random variables and $S_n = X_1 + \cdots + X_n, n \geq 1$. Then for any $\epsilon \geq 0$,

$$P(|S_n| \leq \epsilon \; i.o.) = 0 \text{ or } 1$$

according as $\sum_{n=1}^{\infty} P(|S_n| \leq \epsilon) < \infty$ or $= \infty$.

Solution: Put $C_i = A_i \cap \left(\bigcap_{j=1}^{\infty} A_{i+jk}^c\right), i \geq 1$. Note that $C_i \cap C_{i+jk} = \emptyset$ for $i, j = 1, 2, \ldots$. So for each $n \geq 1$

$$\sum_{i=nk+1}^{\infty} P(C_i) = \sum_{i=n}^{\infty} \sum_{m=1}^{k} P(C_{m+ik})$$

$$= \sum_{m=1}^{k} P\left(\bigcup_{i=n}^{\infty} C_{m+ik}\right) \leq k.$$

Thus for all sufficiently large n,

$$k \geq \sum_{i=nk+1}^{\infty} \left[P(A_i) - P\left(A_i \cap \left(\bigcup_{j=1}^{\infty} A_{i+jk}\right)\right)\right]$$

$$\geq \left(\sum_{i=nk+1}^{\infty} P(A_i)\right)\left(1 - P\left(\bigcup_{j=k}^{\infty} B_j\right)\right)$$

by the given condition.

As the series on the right side is ∞, we must have $P\left(\bigcup_{j=k}^{\infty} B_j\right) = 1$.

(b) It suffices to consider the case when $\sum_{n=1}^{\infty} P(|S_n| \leq \epsilon) = \infty$. To this end, it may be assumed that $\sum_{n=1}^{\infty} P(0 \leq S_n \leq \epsilon) = \infty$, since otherwise $\sum_{n=1}^{\infty} P(-\epsilon \leq S_n \leq 0) = \infty$ and then we can replace X_n by $-X_n$ for each $n \geq 1$.

Put $A_i = [0 \leq S_i \leq \epsilon], B_i = [|S_i| \leq \epsilon], C_{ij} = [|S_j - S_i| \leq \epsilon]$. Note that for each $k \geq 1$,

$$P\left(A_i \cap \left(\bigcup_{j=1}^{\infty} A_{i+jk}\right)\right) \leq P\left(A_i \cap \left(\bigcup_{j=i+k}^{\infty} A_j\right)\right)$$

$$\leq P\left(A_i \cap \left(\bigcup_{j=i+k}^{\infty} C_{ij}\right)\right)$$

$$= P(A_i)P\left(\bigcup_{j=i+k}^{\infty} C_{ij}\right)$$

$$= P(A_i)P\left(\bigcup_{j=i+k}^{\infty} B_{j-i}\right).$$

By (a), $P\left(\cup_{j=k}^{\infty} B_j\right) = 1 \; \forall \, k \geq 1$; i.e., $P(B_n \; i.o.) = 1$.

Example 1.6.14 Let $\{X_n\}$ be iid with the common distribution function F. Let

$$a = \sup\{x \in \mathbb{R} : F(x) < 1\}, \; -\infty < a \leq \infty.$$

(a) If $a < \infty$, show that $\max(X_1, \dots, X_n) \to a \; a.s.$
(b) If $a = \infty$, show that $\max(X_1, \dots, X_n) \to \infty \; a.s.$

Solution: (a) Let $\epsilon > 0$. We shall show that $\sum P(|Y_n - a| > \epsilon) < \infty$ where $Y_n = \max(X_1, \dots, X_n)$. First observe that by definition of a, $P(X_1 \leq a + \epsilon) = 1$. So

$$P(|Y_n - a| < \epsilon) = P(Y_n \leq a - \epsilon) = (P(X_1 \leq a - \epsilon))^n = p^n \text{ (say)}.$$

As $\exists$ a real x such that $F(x) < 1$ and $a - \epsilon < x$, $P(X_1 \leq a - \epsilon) \leq P(X_1 \leq x) < 1$; so $0 \leq p < 1$ and $\sum p^n < \infty$. (b) Let Y_n be as in (a). We show that $\exists \, A \in \mathcal{A}$ such that $P(A) = 0$ and

$$A^c \subset [Y_n \to \infty]. \tag{1.6.3}$$

Firstly, $F(m) < 1 \; \forall \, m \geq 1$ (as $a = \infty$). So $\sum P(Y_n \leq m) = \sum (F(m))^n < \infty$. Then $P(Y_n \leq m \; i.o.(n)) = 0 \; \forall m \geq 1$. Put $A = \cup_{m=1}^{\infty}[Y_n \leq m \; i.o.(n)]$. Then $P(A) = 0$ and $(1.6.3)$ holds; for, $A^c = [\forall m \geq 1, \; Y_n > m \text{ eventually}] \subset [Y_n \to \infty]$.

Example 1.6.15 Let $\{X_n\}$ be pairwise independent, identically distributed and X_1 nondegenerate. Then show that

$$P(\{X_n\} \text{ converges}) = 0.$$

Solution: We show that

$$P(\liminf X_n < \limsup X_n) = 1. \tag{1.6.4}$$

As X_1 is nondegenerate, $\exists \, a \in \mathbb{R}$ such that $0 < P(X_1 \leq a) < 1$ (see Lemma 1.1.3). As $1 > P(X_1 \leq a) = \lim_n P(X_1 \leq a + 1/n)$, $\exists$ an integer $n \geq 1$ such that $P(X_1 \leq a + 1/n) < 1$. So $\exists \, a, b \in \mathbb{R}$ such that $a < b$ and $P(X_1 \leq a) > 0$, $P(X_1 \geq b) > 0$. Thus

$$\sum P(X_n \leq a) = \sum P(X_1 \leq a) = \infty, \; \sum P(X_n \geq b) = \sum P(X_1 \geq b) = \infty.$$

As the X_n are pairwise independent, we must have

$$P(X_n \le a \ i.o.) = 1, \quad P(X_n \ge b \ i.o.) = 1.$$

So
$$P(\liminf X_n \le a) = 1, \quad P(\limsup X_n \ge b) = 1,$$

and hence
$$P(\liminf X_n \le a < b \le \limsup X_n) = 1$$

implying (1.6.4).

Example 1.6.16 Let $\{X_n\}$ be pairwise independent. Show that

$$P(\{X_n\} \text{ converges }) \text{ is } 0 \text{ or } 1.$$

Solution: Let, with the usual convention regarding the empty set,

$$x_0 = \inf\{r \in \mathbb{R} : P(X_n > r \ i.o.) = 0\}, -\infty \le x_0 \le \infty,$$
$$y_0 = \sup\{r \in \mathbb{R} : P(X_n < r \ i.o.) = 0\}, -\infty \le y_0 \le \infty.$$

Then
$$P(\limsup X_n \le x_0) = 1, \quad P(\liminf X_n \ge y_0) = 1, \qquad (1.6.5)$$

(and so $y_0 \le x_0$). We verify the first equality, the proof of the second being similar. To this end, let $x_0 \in \mathbb{R}$ and $k \ge 1$ be an integer. Then $\exists$ an $r_0 \in \mathbb{R}$ such that $r_0 < x_0 + 1/k$ and $P(X_n > r_0 \ i.o.) = 0$. So

$$P(X_n > x_0 + 1/k \ i.o. \ (n)) \le P(X_n > r_0 \ i.o.) = 0;$$

i.e., $P(X_n \le x_0 + 1/k$ eventually $(n)) = 1$ implying that

$$P(\limsup X_n \le x_0 + 1/k) = 1.$$

As $k \ge 1$ is arbitrary, we can conclude that $P(\limsup X_n \le x_0) = 1$. It, therefore, remains to consider the case $x_0 = -\infty$. Let $\gamma_m \to -\infty$ be such that $P(X_n > \gamma_m \ i.o.(n)) = 0 \ \forall \ m \ge 1$. So

$$P(X_n \le \gamma_m \text{ eventually } (n) \ \forall \ m \ge 1) = 1$$

which implies that $P(\limsup X_n = -\infty) = 1$.

Clearly, (1.6.5) implies that if $x_0 = -\infty$ or $y_0 = +\infty$, then $P(\{X_n\} \text{ converges})$ $= 0$. Now, let $y_0 = x_0 \in \mathbb{R}$; then (1.6.5) implies that $P(\{X_n\} \text{ converges}) = 1$. Finally, assume that $y_0 < x_0$. Then $\exists$ reals u, v such that $y_0 < u < v < x_0$. By the definitions of x_0 and y_0, it then follows that

$$P(X_n < u \ i.o.) > 0, \ P(X_n > v \ i.o.) > 0.$$

By the Borel zero-one law (see Theorem 3.1.2), we have

$$P(X_n < u \ i.o.) = 1, \ P(X_n > v \ i.o.) = 1;$$

i.e.,

$$P(\liminf X_n \leq u) = 1, \ P(\limsup X_n \geq v) = 1.$$

As $u < v$, we must have $P(\liminf X_n < \limsup X_n) = 1$. i.e.,

$$P(\{X_n\}\text{converges}) = 0.$$

[Under the given conditions, we, therefore, have

$$P(\lim X_n \text{ exists in the extended real number system}) = 0 \text{ or } 1.]$$

Example 1.6.17 (Due to D.J. Newman; see Feller (1968, p. 210)) Let $\{X_n\}_{n \geq 1}$ be a sequence of independent Bernoulli variables with $P(X_n = 1) = p = 1 - P(X_n = 0)$. Define Y_n to be the length of the maximal run of successes starting at the nth trial:

$$Y_n(\omega) = j \text{ iff } X_i(\omega) = 1 \text{ for } i = n, \ldots, n + j - 1$$
$$\text{and } X_{n+j}(\omega) = 0 \ (j \geq 1),$$
$$Y_n(\omega) = 0 \text{ iff } X_n(\omega) = 0.$$

Let log n stand for the logarithm of n to the base $1/p, 0 < p < 1$. Then show that $\limsup(Y_n/\log n) = 1 \ a.s.$ (See, also, Example 3.1.1 on p. 64.)
Solution: For each $a > 1$, we have

$$P(Y_n > a \log n) = \sum_{j=m}^{\infty} (1 - p)p^j = p^m \leq p^{a \log n} = n^{-a}$$

where $m = \inf\{j \geq 1 : j > a \log n\}$. So $P(Y_n > a \log n \ i.o.) = 0$. Since $a > 1$ is arbitrary, we must have

$$P(\limsup(Y_n/\log n) \leq 1) = 1.$$

Now let $j_n = [n \log n]$, the greatest integer $\leq n \log n$. Put

$$\log_2 n = \log \log n.$$

Let $0 < b < 1$. Note that

$$
\begin{aligned}
j_{n+1} - j_n &\geq (n+1)\log n - 1 - n\log n \\
&\geq [b\log j_n] + \log n - 1 - b(\log n + \log_2 n) \\
&= [b\log j_n] + (1-b)\log n\left[1 - \frac{1 + b\log_2 n}{(1-b)\log n}\right].
\end{aligned}
$$

The second term on the right side above tends to ∞ as $n \to \infty$, so that $\exists$ an integer $m \geq 1$ such that it is > 1 whenever $n \geq m$. We assume below that $n \geq m$. Put $j_n^* = j_n + [b\log j_n]$. Note that

$$
A_n := [Y_{j_n} > j_n^* - j_n + 1] = [X_{j_n} = 1, \ldots, X_{j_n^*} = 1]
$$

which depends only on $X_{j_n}, \ldots, X_{j_n^*}$; this immediately implies that the events $\{A_n\}_{n\geq 1}$ are independent. Also,

$$
P(A_n) = p^{[b\log j_n]+1} \geq p(n\log n)^{-b}.
$$

An application of the second Borel–Cantelli lemma yields that

$$
P(A_n \; i.o.) = 1,
$$

and hence that $P(\limsup[Y_n > b\log j_n]) = 1$. As $b \in (0,1)$ is arbitrary, we get

$$
\begin{aligned}
&P(\limsup(Y_n/\log n) \geq 1) \\
&\geq P(\limsup(Y_{j_n}/\log j_n) \geq 1) = 1.
\end{aligned}
$$

Example 1.6.18 Let $\{X_n\}_{n\geq 1}$ be pairwise independent and identically distributed with $E(|X_1|) = \infty$. Let $S_n = X_1 + \cdots + X_n, n \geq 1$. Let $\{a_n\}_{n\geq 1}$ be a sequence of positive reals such that $n^{-1}a_n$ is nondecreasing. Then

$$
\sum P(|X_1| > a_n) = \infty \Rightarrow \limsup(|S_n|/a_n) = \infty \; a.s.
$$

Solution: As $a_{kn}/(kn) \geq a_n/n \; \forall k \geq 1$, we must have $a_{kn} \geq k a_n \; \forall n, k \geq 1$. Hence

$$
\sum_{n=1}^{\infty} P(|X_1| > ka_n) \geq \sum_{n=1}^{\infty} P(|X_1| > a_{kn}) \geq k^{-1}\sum_{m=k}^{\infty} P(|X_1| > a_m) = \infty
$$

since

$$
\sum_{m=kn}^{k(n+1)} P(|X_1| > a_m) \geq \sum_{m=kn}^{k(n+1)} P(|X_1| > a_{kn}) = k P(|X_1| > a_{kn}).
$$

Thus $\sum P(|X_n| \geq k\, a_n) = \infty$, and so $P(|X_n| \geq k\, a_n \ i.o.) = 1$. Proceeding as in Example 1.6.6, one gets the desired result.

Example 1.6.19 Let $\sum P(X_n \neq Y_n) < \infty$. Then

(a) $\sum (X_n - Y_n)Z_n$ converges a.s.;
(b) if $a_n \to \infty$ and $\{m_n\}_{n \geq 1}$ is a sequence of positive integers tending to $+\infty$,

$$\frac{1}{a_n} \sum_{i=1}^{m_n} (X_i - Y_i)Z_i \to 0 \ a.s.;$$

(c) $(X_n - Y_n)Z_n \to 0 \ a.s.$;
(d) with probability one,

$$\sum X_n Z_n \ \text{or} \ \frac{1}{a_n} \sum_{i=1}^{n} X_i Z_i \ \text{or} \ X_n Z_n$$

converges, tends to $+\infty$ or $-\infty$, or fluctuates in the same way as

$$\sum Y_n Z_n \ \text{or} \ \frac{1}{a_n} \sum_{i=1}^{n} Y_i Z_i \ \text{or} \ Y_n Z_n$$

, respectively, where $a_n \to +\infty$;
(e) $\sum_{i=1}^{n} X_i / a_n \to^P X \Leftrightarrow \sum_{i=1}^{n} Y_i / a_n \to^P X$.

Solution: By the first Borel–Cantelli lemma, $P(X_n \neq Y_n \ i.o.) = 0$. Thus $P(A) = 1$ where $A = \lim \inf [X_n = Y_n]$. Now note that

$$A \subset [\sum (X_n - Y_n)Z_n \ \text{converges}],$$

$$A \subset \left[\sum_{i=1}^{m_n} (X_i - Y_i)Z_i / a_n \to 0 \right]$$

$$A \subset [(X_n - Y_n)Z_n \to 0].$$

Parts (d) and (e) follow from Parts (a)–(c).

Example 1.6.20 Let X be a random variable satisfying the condition that $\exists$ a sequence $\{d_n\}_{n \geq 1}$ of non-negative reals such that $P(|X| \geq d_n) > 0\, \forall n \geq 1$ and $\sum_{n=1}^{\infty} P(|X| \geq d_n) < \infty$. Let $\{a_n\}_{n \geq 1}$ be a given sequence of reals. Then $\exists$ a sequence $\{X_n\}_{n \geq 1}$ of random variables such that $X_n \to X \ a.s.$ and $E(X_n) = a_n \ \forall n \geq 1$.
Solution: Define

$$X_n = X I_{[|X| < d_n]} + \frac{a_n - \alpha_n}{P(|X| \geq d_n)} I_{[|X| \geq d_n]}, \ n \geq 1,$$

where $\alpha_n = E(X I_{[|X|<d_n]})$; note that the random variable $X I_{[|X|<d_n]}$ is bounded (by d_n) and hence has a finite expectation.

Clearly, $E(X_n) = a_n, n \geq 1$. Also, $X_n \to X$ a.s., since for any $\epsilon > 0$

$$\sum P(|X_n - X| > \epsilon) \leq \sum P(|X| \geq d_n) < \infty.$$

Example 1.6.21 Let $P(X_n \geq x) \leq P(Y \geq x) \, \forall x > 0, n \geq 1$. Let $Y_n = \max(X_1, \ldots, X_n), n \geq 1$. If $E(Y^+) < \infty$, then $n^{-1} Y_n \to 0$ a.s.

Solution: Note that if $\epsilon > 0$,

$$\sum P(X_n \geq n\epsilon) \leq \sum P(Y \geq n\,\epsilon) \leq E(Y^+/\epsilon) < \infty \text{ by Lemma 1.1.1 (b).}$$

So $P(X_n \geq n\,\epsilon \ i.o.) = 0$. Hence $P(A) = 1$ where

$$A = [X_n < n/m \, i.o.(n) \, \forall m \geq 1].$$

We now show that

$$A \subset \left[\limsup(n^{-1}Y_n) \leq 0 \right].$$

To this end, let $\omega \in A$. Let $m \geq 1$. Then $\exists$ an integer $N(\omega) \geq 1$ such that $X_n < n/m \, \forall n \geq N(\omega)$, and so

$$n^{-1}Y_n(\omega) \leq \max(n^{-1}Y_{N(\omega)}(\omega), 1/m) \, \forall n \geq N(\omega)$$

which implies that

$$\limsup(n^{-1}Y_n(\omega)) \leq 1/m.$$

As $m \geq 1$ is arbitrary, we must have $\limsup(n^{-1}Y_n(\omega)) \leq 0$. Thus $\limsup(n^{-1}Y_n) \leq 0$ a.s. But

$$\liminf(n^{-1}Y_n) \geq \liminf(n^{-1}X_1) = 0.$$

So $n^{-1}Y_n \to 0$ a.s.

Example 1.6.22 (a) If $\sum P(X_n \geq A) < \infty$ for some real A, then

$$P\left(\sup_n X_n < \infty \right) = 1.$$

(b) If the events $[X_n > A]$ are pairwise NQD and $\sum P(X_n > A) = \infty$ for each real A, then $P\left(\sup_n X_n < \infty \right) = 0$.

Solution: This is immediate from the Borel–Cantelli lemmas.

Example 1.6.23 (a) If $P(A_n) \to 1$, then $\exists$ a subsequence $\{n_k\}_{k\geq 1}$ of positive integers such that $P\left(\cap_{k=1}^{\infty} A_{n_k}\right) > 0$.
(b) The sufficient condition of (a) cannot be replaced by '$P(A_n) \geq \epsilon \; \forall n \geq 1$ for some $\epsilon > 0$'.

Solution: (a) As $P(A_n^c) \to 0$, $\exists$ a subsequence $\{m_k\}_{k\geq 1}$ of positive integers such that $P(A_{m_k}^c) \leq k^{-2} \forall k \geq 1$. Then $P(\liminf A_{m_k}) = 1$. Thus

$$P\left(\overset{\infty}{\underset{j=k}{\cap}} A_{m_j}\right) \to 1 \text{ as } k \to \infty.$$

Hence $\exists k_0 \geq 1$ such that $P\left(\cap_{j=k_0}^{\infty} A_{m_j}\right) > 0$. Then we consider the subseuence $m_{k_0}, m_{k_0+1}, m_{k_0+2}, \ldots$

(b) Let us consider the experiment of independent coin-tossing with the same coin. Let $A_n =$ [a Head appears at the n-th toss], $n \geq 1$. Then the events A_n are independent and $P(A_n) = 1/2 \; \forall \; n \geq 1$. So we can take $\epsilon = 1/2$. Yet, for any subsequence $n_1 < n_2 < \cdots$, we have

$$P\left(\overset{\infty}{\underset{k=1}{\cap}} A_{n_k}\right) \leq P\left(A_{n_1} \cap \cdots \cap A_{n_k}\right) \forall k \geq 1$$
$$= \left(\frac{1}{2}\right)^k \forall k \geq 1,$$

which implies that $P\left(\cap_{k=1}^{\infty} A_{n_k}\right) = 0$.

Example 1.6.24 In a sequence of iid Bernoulli random variables $\{X_n\}_{n\geq 1}$ with $P(X_1 = 1) = p$, let A_n be the event that a run of n consecutive 1's occurs between the 2^n-th and 2^{n+1}-th trials, $n \geq 1$. If $p \geq 1/2$, then $P(A_n \; i.o.) = 1$.

Solution: Clearly, the events A_n are independent. So it suffices to show that $\sum P(A_n) = \infty$. We now show that

$$P(A_n^c) \leq \exp(-(2p)^n/(2n)), n \geq 1.$$

To this end, note that

$$A_n^c = [\text{ for each } i = 2^n, 2^n + 1, \ldots, 2^{n+1} - n + 1, \text{ there is at least}$$
$$\text{one zero between the } i\text{-th and } (i + n - 1)\text{-th trials }]$$
$$\subset [\text{for each } i = 2^n, 2^n + n, 2^n + 2n, \ldots, 2^n + rn, \text{ there is}$$
$$\text{at least one zero between the } i\text{-th and } (i + n - 1)\text{-th trials }]$$

where r is the largest integer such that $2^n + rn \geq 2^{n+1} - n + 1$, and so $r \geq (2^n - n + 1)/n \geq 2^n/(2n)$—note that $2^{n-1} \geq n - 1$ for $n \geq 1$ as an induction on n

shows. Hence

$$P(A_n^c) \leq (1 - p^n)^r \leq (1 - p)^{2^n/(2n)} \leq \exp(-(2p)^n/(2n)).$$

Case 1 $p = 1/2$

Then $P(A_n) \geq (1/(2n))/(1 + 1/(2n))$ so that $\sum P(A_n) = \infty$.

Case 2 $p > 1/2$

Then $P(A_n^c) \to 0$ so that $\sum P(A_n) = \infty$.

Example 1.6.25 (a) Let $Y_n = \max\{X_1, \ldots, X_n\}, n \geq 1$ and $\lambda_n \uparrow \infty$. Then $[Y_n > \lambda_n \ i.o.] = [X_n > \lambda_n \ i.o.]$

(b) If $\sum P(X_n > \lambda_n) < \infty$, then $P(Y_n > \lambda_n \ i.o.) = 0$; if $\sum P(X_n > \lambda_n) = \infty$ and the events $[X_n > \lambda_n]$ are pairwise NQD, then $P(Y_n > \lambda_n \ i.o.) = 1$.

Solution: (a) If suffices to show that

$$[Y_n > \lambda_n \ i.o.] \subset [X_n > \lambda_n \ i.o.].$$

So let $Y_n(\omega) > \lambda_n$ for $n = n_1, n_2, \ldots$ where $n_1 < n_2 < \ldots$ Then $\exists$ an integer k such that $1 \leq k \leq n_1$ and $X_k(\omega) > \lambda_{n_1}$, and so $X_k(\omega) > \lambda_k$ (as $\lambda_k \leq \lambda_{n_1}$). Put $m_1 = k$.
Suppose that $\exists m_1 < m_2 < \cdots < m_j$ such that

$$X_n(\omega) > \lambda_n \text{ for } n = m_1, \ldots, m_j.$$

Then $\exists$ an integer n_i such that $\lambda_{n_i} \geq Y_{m_j}(\omega)$ (this is possible since $\lambda_n \to \infty$). As $Y_{n_i}(\omega) > \lambda_{n_i}$, there must be an integer $p \geq 1$ such that $p \leq n_i$ and $X_p(\omega) > \lambda_{n_i}$. Clearly, $p > m_j$. We can let $m_{j+1} = p$. Therefore, by mathematical induction, $\exists m_1 < m_2 < \ldots$ such that $X_n(\omega) > \lambda_n$ for $n = m_1, m_2, \ldots$
(b) It is now immediate from (a).

Example 1.6.26 Let $\{f_n\}_{n \geq 1}$ be q sequence of functions from Ω into $\mathbb{R}$. Let $b_n > 0 \, \forall \, n \geq 1$.

(a) If $\liminf(a_n/b_n) > 1$, then

$$[f_n \geq a_n \ i.o.] \subset [f_n > b_n \ i.o.].$$

(b) If $\limsup(a_n/b_n) < 1$, then

$$[f_n \leq a_n \ i.o.] \subset [f_n < b_n \ i.o.].$$

Solution: This is immediate.
    Additional problems on BCL can be found in Athreya and Lahiri (2006, p. 43) and Stein and Shakarchi (2005, p. 46).

References

B.C. Arnold, Some elementary variations of the Lyapounov inequality. SJAM **35**, 117–118 (1978)

R.B. Ash, C.A. Doléans-Dade, *Probability and Measure Theory* (Academic Press, Second Edition, (2000)

K.B. Athreya, S.N. Lahiri, *Probability Theory*, Trim Series 41 (Hindustan Book Agency, India, 2006)

P. Billingsley, *Probability and Measure*, 3rd edn. (Wiley, New York, 1995). Second Edition 1991. First Edition 1986

É. Borel, Les probabilités dénombrables et leurs applications arith-métiq-ues. Rend. Circ. Mat. Palermo **27**, 247–271 (1909)

É. Borel, Sur un problème de probabilités relatif aux fractions continues. Math. Ann. **77**, 578–587 (1912)

É. Borel, Traité du calcul des probabilitités et de ses applications, 2, No.1, Applications d l' arith-métique et à la théorie des fonctions. Gauthier-Villars, Paris (1926)

L. Breiman, *Probability* (Addision Wesley, California, 1968)

F.P. Cantelli, Sulla probabilità come limite della frequenza. Rend. Accad. Lincai Ser.5. 24, 39–45 (1917)

T.K. Chandra, *A First Course in Asymptotic Theory of Statistics* (Narosa Publishing House Pvt. Ltd., New Delhi, 1999)

Y.S. Chow, H. Teicher, *Probability Theory* (Springer, New York, 1997)

K.L. Chung, *A Course in Probability Theory* (Academic Press, New York, 2001)

K.L. Chung, P. Erdös, On the application of the Borel-Cante!!i lemma. TAMS **72**, 179–186 (1952)

D.A. Dawson, D. Sankoff, An inequality of probabilities. PAMS **18**, 504–507 (1967)

B. Eisenberg, B.K. Ghosh, A generalization of Markov's inequality. SPL **53**, 59–65 (2001)

P. Erdös, A. Renyi, On Cantor's series with convergent $\sum 1/q_n$. Ann. Univ. Sci. Budapest Eötvós Sect. Math. **2**, 93–109 (1959)

W. Feller, *An Introduction To Probability Theory And Its Applications*, vol. I, 3rd edn. (Revised), (Wiley, New York, 1968)

C. Feng, L. Li, J. Shen, On the Borel-Cantelli lemma and its generalization. C.R. Acad. Sci. Paris Ser. I **347**, 1313–1316 (2009)

A. Gut, *Probability: A Graduate Course* (Springer, New York, 2005)

S. Kochen, C. Stone, A note on the Borel-Cantelli lemma. IJM **8**, 248–251 (1964)

A.N. Kolmogorov, *Grundbegriffe der Wahrscheinlickheitsrechnung*, Berlin. Foundation of the theory of Probability (English Translation), 1956, 2nd edn. (Chelsea, New York, 1933)

J. Lamperti, Wiener's test and Markov chains. J. Math. Anal. Appl. **6**, 58–66 (1963)

M. Loève, *Probability Theory*, 4th edn. (Springer-Verlag, New York, 1977)

T.F. Móri, G.J. Székeley, On the Erdös-Rényi generalization of the Borel-Cantelli lemma. Studia Sci. Math. Hung. **18**, 173–182 (1983)

S.W. Nash, An extension of the Borel-Cantelli lemma. AMS **25**, 165–167 (1954)

R.E.A.C. Paley, A. Zygmund, On some sequences of functions III. Proc. Camb. Phil. Soc. **28**, 190–205 (1932)

V.V. Petrov, *Sums of Independent Random Variables* (Springer-Verlag, New York, 1975a)

V.V. Petrov, An inequality for moments of a random variable. TPA **20**, 391–392 (1975b)

V.V. Petrov, *Limit Theorems of Probability Theory* (Oxford University Press, New York, 1995)

V.V. Petrov, On lower bounds for tail probabilities. J. Statist. Plan Inference **137**, 2703–2705 (2007a)

V.V. Petrov, A generalization of the Chung-Erdös inequality for the probability of a union of events. JMS **147**, 6932–6934 (2007b)

H.L. Royden, *Real Analysis*, 3rd edn. (Macmillan Publishing Company, New York, 1988)

L.A. Rubel, A complex-variables proof of hölder's inequality. PAMS **15**, 999 (1964)

W. Rudin, *Real and Complex Analysis*, 3rd edn. (McGraw-Hill Book Company, New York, 1987)

J. Shuster, On the Borel-Cantelli problem. Can. Math. Bull. **13**, 273–275 (1970)

E.M. Stein, R. Shakarchi, *Real Analysis* (Princeton University Press, Princeton, 2005)

Chapter 2
Extensions of the First BCL

2.1 A Result of Barndorff-Nielsen

The first Borel–Cantelli lemma is simple and almost trivial. Yet, it is necessary to weaken its sufficient condition to tackle some problems of probability theory. The first of such extensions is due to Barndorff-Nielsen (1961), and will be stated below.

Theorem 2.1.1 *Let* $\{A_n\}_{n\geq 1}$ *be a sequence of events such that* $\liminf P(A_n) = 0$ *and* $\sum P(A_n \cap A_{n+1}^c) < \infty$. *Then*

$$P(\limsup A_n) = 0 \text{ and } P(A_n) \to 0.$$

Proof Put $B_n = A_n \cap A_{n+1}^c, n \geq 1$. Then $B_n \in \mathcal{A} \forall n \geq 1$, and $P(\limsup B_n) = 0$ by the first Borel–Cantelli lemma. By (1.2.12) on page 8,

$$(\limsup A_n) \cap (\limsup A_n^c) \subset \limsup B_n.$$

Therefore, inequality (h) of Sect. 1.1 on page 2 implies that

$$P(\limsup A_n) \leq P(\limsup B_n) + P(\liminf A_n).$$

This completes the proof, since $P(\liminf A_n) = 0$ by (1.2.4) on page 7. □
The above proof suggests the following extension of Theorem 2.1.1.

Theorem 2.1.2 *Assume that*

(a) $P(\liminf A_n) = 0$ *(a fortiori,* $\liminf P(A_n) = 0$*); and*
(b) $P(\limsup(A_n \cap A_{n+1}^c)) = 0$ *or* $P(\limsup(A_n^c \cap A_{n+1})) = 0.$

Then $P(\limsup A_n) = 0.$ □

The paper by Barndorff-Nielsen (1961) contains an application of Theorem 2.1.1. It is worthwhile to state the following analog of Theorem 2.1.1.

T. K. Chandra, *The Borel–Cantelli Lemma*, SpringerBriefs in Statistics,
DOI: 10.1007/978-81-322-0677-4_2, © The Author(s) 2012

Theorem 2.1.1′ If $\liminf P(A_n) = 0$ and $\sum_n P(A_n^c \cap A_{n+1}) < \infty$, then

$$P(\limsup A_n) = 0. \qquad \qquad \square$$

There is a short proof of the above result which is due to Balakrishnan and Stepanov (2010). This runs as follows: For each $n \geq 1$,

$$P(\limsup A_n) \leq P\left(\bigcup_{m=n}^{\infty} A_m\right) = P(A_n) + P(A_{n+1} \cap A_n^c)$$
$$+ P(A_{n+2} \cap A_{n+1}^c \cap A_n^c) + \cdots$$
$$\leq P(A_n) + \sum_{i=n}^{\infty} P(A_i^c \cap A_{i+1}).$$

As $\sum_n P(A_n^c \cap A_{n+1}) < \infty$, $\displaystyle\lim_{n\to\infty} \sum_{i=n}^{\infty} P(A_i^c \cap A_{i+1}) = 0$. So

$$P(\limsup A_n) \leq \liminf P(A_n) + \lim_{n\to\infty} \sum_{i=n}^{\infty} P(A_i^c \cap A_{i+1}) = 0.$$

Replacing A_n by $A \cap A_n$ for $n \geq 1$ in Theorem 2.1.2, and using

$$P(\limsup A_n) \leq P((\limsup A_n) \cap A) + P(A^c),$$

we get a further refinement of Theorem 2.1.2.

Theorem 2.1.3 *Assume that*

(a) $P(\liminf(A \cap A_n)) = 0$ (a fortiori, $\liminf P(A \cap A_n) = 0$); and
(b) $P(\limsup(A \cap A_n \cap A_{n+1}^c)) = 0$ or $P(\limsup(A \cap A_n^c \cap A_{n+1})) = 0$.

Then $P(\limsup A_n) \leq 1 - P(A)$, $\limsup P(A_n) \leq 1 - P(A)$. $\qquad \square$

Theorem 2.1.4 (Balakrishnan and Stepanov 2010) *If $P(A_n) \to 0$ and*

$$\sum_{n=1}^{\infty} P(A_n^c \cap A_{n+1}^c \cap \cdots \cap A_{n+m-1}^c \cap A_{n+m}) < \infty \qquad (2.1.1)$$

for some $m \geq 1$, then $P(A_n \ i.o.) = 0$.

Proof Note that for each $n \geq 1$,

$$P(A_n \ i.o.) \le P\left(\overset{\infty}{\underset{k=n}{\cup}} A_k\right)$$

$$= P(A_n) + P(A_n^c \cap A_{n+1}) + P(A_n^c \cap A_{n+1}^c \cap A_{n+2}) + \cdots$$

$$\le P(A_n) + P(A_{n+1}) + \cdots + P(A_{n+m-1})$$

$$+ \sum_{k=n}^{\infty} P(A_k^c \cap A_{k+1}^c \cap \cdots \cap A_{k+m-1}^c \cap A_{k+m})$$

$$\to 0 \text{ as } n \to \infty. \qquad \qquad \square$$

The proof of Theorem 2.1.4 shows that

$$P(A_n \ i.o.) = \lim_{n \to \infty} \left[P(A_n) + \sum_{k=1}^{\infty} P(A_n^c \cap A_{n+1}^c \cap \cdots \cap A_{n+k-1}^c \cap A_{n+k}) \right].$$

There is a dual of Theorem 2.1.4. Before stating it, we shall give an alternative proof of Theorem 2.1.4. The rest of this section is due to Riddhipratim Basu, a graduate student of the Indian Statistical Institute.

Lemma 2.1.1 *Let for some* $m \ge 1$,

$$C_n = \overset{m}{\underset{j=1}{\cup}} A_{n+j},$$
$$B_n = A_n^c \cap A_{n+1}^c \cap \cdots \cap A_{n+m-1}^c \cap A_{n+m}, n \ge 1.$$

Then

$$\limsup A_n \subset (\limsup B_n) \cup (\liminf C_n).$$

Proof Let ω lie in LHS. Then $\omega \in A_n$ for infinitely many values of n. Let

$$\{n \ge 1 : \omega \in A_n\} = \{n_1, n_2, \ldots\} \quad \text{where } n_1 < n_2 < \cdots$$

Let $k = \limsup_{i \to \infty} \{n_i - n_{i-1}\}$.

Case 1 $k \ge m + 1$.

Then $\exists$ a subsequence $\{n_{i_j}\}_{j \ge 1}$ of $\{n_i\}_{i \ge 1}$ such that $n_{i_1} \ge m+1$ and $n_{i_j} - n_{i_j-1} \ge m + 1 \forall j \ge 1$; one can verify this by considering the three subcases '$k = \infty$', '$k = m + 1$', and '$k \ge m+2$'. Then $\omega \in B_{n_{i_j}-m}, \forall j \ge 1$; for, if we fix a $j \ge 1$, $\omega \in A_{n_{i_j}}$, and as $n_{i_j-1} \le n_{i_j} - m - 1$, we get by the definition of $\{n_i\}_{n \ge 1}$

$$\omega \notin A_{n_{i_j}-m}, \omega \notin A_{n_{i_j}-m+1}, \ldots, \omega \notin A_{n_{i_j}-1}.$$

So $\omega \in \limsup B_n$.

Case 2 $1 \le k \le m$.

Then $\limsup_{i\to\infty}\{n_i - n_{i-1}\} < m + \dfrac{1}{2}$ so that $\exists i_0 \geq 1$ such that $n_{i_0} \geq m + 1$ and $n_i - n_{i-1} \leq m \forall i \geq i_0$. But then

$$\omega \in C_{n-m} \forall n \geq n_{i_0}.$$

To verify this, fix an integer $n \geq n_{i_0}.\exists j \geq 1$ such that $n_j \leq n < n_{j+1}$ for some $j \geq i_0$. Now note that that $\omega \in A_{n_j}$, and so $\omega \in C_{n-m}$ since $n - m + 1 \leq n_{j+1} - 1 - m + 1 \leq n_j$ which implies that $n_j \in \{n - m + 1, \ldots, n\}$ and so $C_{n-m} \supset A_{n_j}$.

Hence $\omega \in \liminf C_n$. □

The above lemma immediately implies the following result.

Theorem 2.1.4′ If $P(\liminf C_n) = 0$ and for some $m \geq 1$, (2.1.1) holds, then

$$P(A_n \ i.o.) = 0,$$

where C_n is as in Lemma 2.1.1. □

Lemma 2.1.2 Let C_n be as in Lemma 2.1.1, and let

$$B_n^* = A_n \cap A_{n+1}^c \cap \cdots A_{n+m}^c \ for \ some \ m \geq 1.$$

Then

$$\limsup A_n \subset (\limsup B_n^*) \cup (\liminf C_n).$$

Proof Let $\omega \in \limsup A_n$ and $\omega \notin \limsup B_n^*$. Then $\exists$ an integer $m \geq 1$ such that $\omega \notin B_n^* \forall n > m$. Also, $\exists n_0 > m$ such that $\omega \in A_{n_0}$ and so $\omega \notin C_{n_0-1}$. We assert that $\omega \notin C_n \forall n \geq n_0 - 1$. Suppose this is false. Then $\exists$ an integer $n_1 \geq n_0$ such that $\omega \notin C_{n_1}$. Hence

$$\omega \notin A_{n_1+1}, \ldots, \omega \notin A_{n_1+m}.$$

Let n_2 be the largest integers less than $n_1 + 1$ such that $\omega \in A_{n_2}$. Then

$$\omega \in A_{n_2+1}^c, \omega \in A_{n_2+2}^c, \ldots, \omega \in A_{n_2+m}^c.$$

(Distinguish between two cases, e.g., $n_2 + m \leq n_1$ or $n_2 + m > n_1$; in the latter case, note that $n_2 + m \leq n_1 + m$.) Thus $\omega \in B_{n_2}^*$ but $n_2 \geq n_0 > m$. This is a contradiction. □

We have thus proved.

Theorem 2.1.5 If $P(\liminf C_n) = 0$ and for some $m \geq 1$,

$$\sum P(B_n^*) < \infty,$$

then $P(A_n \ i.o.) = 0$ *where* C_n *and* B_n^* *are as in Lemma 2.1.2.*   □

We can combine Theorems 2.1.4′ and 2.1.5 in the following way:

Theorem 2.1.6 *If* $P(\liminf C_n) = 0$ *and*

$$P(\limsup B_n) = 0 \ or \ P(\limsup B_n^*) = 0,$$

then $P(A_n \ i.o.) = 0$ *where* C_n, B_n *and* B_n^* *are as in Lemmas 2.1.1. and 2.1.2.*   □

Remark 2.1.1 This remark is related to Theorems 2.1.4 and 2.1.5. Suppose that $P(A_n) \to 0$ and

$$\sum P(B_n \cap B_{n+1} \cap \cdots \cap B_{n+m}) < \infty$$

where each B_i is either A_i and A_i^c and at least two of the B_i for $i = n, \ldots, n + m$ are the corresponding A_i. Then it need not true that $P(A_n \ i.o.) = 0$. For, we have the counterexample : Let $\{A_n\}_{n\geq 1}$ be independent and $P(A_n) = \frac{1}{n}, n \geq 1$; then the above conditions hold, but $P(A_n \ i.o.) = 1$.

We now give two applications of Theorem 2.1.1.

Example 2.1.1 Let $\{X_n\}_{n\geq 1}$ be pairwise independent, and assume that for each $n \geq 1$,

$$P(X_n > u) = e^{-u}, \quad 0 < u < \infty.$$

(a) Show that

$$\limsup(X_n / \log n) = 1 \ a.s. \ and \ \liminf(X_n / \log n) = 0 \ a.s. \qquad (2.1.2)$$

(b) Let $X_{(n)} = \max(X_1, \ldots, X_n), n \geq 1$. If $\{X_n\}_{n\geq 1}$ is independent, then

$$X_{(n)} / \log n \to 1 \ a.s.$$

Solution: (a) By the Borel-Cantelli lemmas, we have

$$P(X_n > a \log n \ i.o.) = 0 \ \text{or} \ 1 \ \text{according as} \ a > 1 \ \text{or} \ 0 < a \leq 1.$$

This implies the first part of (2.1.2). Since for any $a > 0$, $P(X_n < a \log n) \to 1$, and so $\sum P(X_n < a \log n) = \infty$, $P(X_n < a \log n \ i.o.) = 1 \forall a > 0$. Therefore,

$$\liminf(X_n / \log n) \leq 0 \ a.s.$$

which is tantamount to the second part of (2.1.2).

(b) By Example 1.6.25 (a) on page 48, the above arguments yield that $P(X_{(n)} > a \log n \ i.o.) = 0$ and hence

$$\limsup(X_{(n)} / \log n) \leq 1 \ a.s.$$

We show below that $\liminf(X_{(n)}/\log n) \geq 1$ $a.s.$ For this, it suffices to show that $P(X_{(n)} > a \log n$ eventually$) = 1$ for each $a \in (0, 1)$. By Theorem 2.1.2, it is enough to show that, for $0 < a < 1$,

$$\sum P(A_n \cap A_{n+1}^c) < \infty \text{ and } P(A_n^c \text{ } i.o.) = 1, \tag{2.1.3}$$

where $A_n = [X_{(n)} \leq a \log n]$, $n \geq 1$. By Example 1.6.25 (a), we have

$$P(A_n^c \text{ } i.o) = P(X_n > a \log n \text{ } i.o.) = 1 \text{ for } 0 < a < 1.$$

Finally,

$$\begin{aligned}
\sum_{n=1}^{\infty} P(A_n \cap A_{n+1}^c) &= \sum_{n=1}^{\infty} P(A_n \cap [X_{n+1} > a \log(n+1)]) \\
&= \sum_{n=1}^{\infty} P(A_n) P(X_{n+1} > a \log(n+1)) \\
&= \sum_{n=1}^{\infty} (1 - n^{-a})^n (n+1)^{-a} \\
&\leq \sum_{n=1}^{\infty} n^{-a} \exp(-n^{1-a}) \\
&\leq \sum_{n=1}^{\infty} \int_{n-1}^{n} x^{-a} \exp(-x^{1-a}) dx \quad \text{(why?)} \\
&= \int_0^{\infty} x^{-a} \exp(-x^{1-a}) dx \\
&= \int_0^{\infty} e^{-y} dy/(1-a) = 1/(1-a) < \infty.
\end{aligned}$$

Example 2.1.2 Let $\{X_n\}$ be pairwise independent and each X_n follow $N(0; 1)$ distribution.

(a) Show that

$$\limsup(X_n/(\sqrt{2\log n})) = 1 \text{ } a.s., \text{ and } \liminf(X_n/(\sqrt{2\log n})) = -1 \text{ } a.s. \tag{2.1.4}$$

(b) Show that, if $X_{(n)} = \max(X_1, \ldots, X_n)$, $n \geq 1$ and $\{X_n\}_{n \geq 1}$ is independent then

$$X_{(n)}/\sqrt{2\log n} \to 1 \text{ } a.s.$$

Solution: We shall use the inequality (1.8) on page 175 of Feller (1968).

(a) Since $P(N(0; 1) > x) \leq \frac{1}{\sqrt{2\pi}} \exp(-\frac{1}{2}x^2)$ for $x \geq 1$, we have

$$P(X_n > a\sqrt{2\log n}) = 0 \text{ or } 1 \text{ according as } a > 1 \text{ or } 0 < a \leq 1$$

which implies the first part of (2.1.3). Replacing X_n by $-X_n$ for each $n \geq 1$, we get the second part of (2.1.3).

(b) Following the steps of Example 2.1.1 (b), it suffices to show that for each $a \in (0, 1)$

$$\sum P(A_n \cap A_{n+1}^c) < \infty$$

where $A_n = [X_{(n)} \leq a\sqrt{2\log n}]$, $n \geq 1$. To do this, note that for a suitable $m \geq 1$

$$\sum_{n=m}^{\infty} P(A_n \cap A_{n+1}^c) \leq \sum_{n=m}^{\infty} \left(1 - \frac{1}{2a\sqrt{2\pi}\sqrt{2\log n}} \exp(-a^2 \log n)\right)^n$$

$$\times \frac{1}{\sqrt{2\pi}} \exp(-a^2 \log(n+1)) \tag{2.1.5}$$

$$\leq d \sum_{n=m}^{\infty} n^{-a^2} \exp(-cn^{-a^2+1}/\sqrt{\log n})$$

where $c > 0$, $d > 0$ are suitable constants. In (2.1.5), we have used the fact that

$$P(N(0; 1) > x) \geq \frac{1}{\sqrt{2\pi}} \exp(-\frac{1}{2}x^2) \left(\frac{1}{x} - \frac{1}{x^3}\right) \quad \text{for } x \geq 1$$

and m is an integer such that $(1 - 1/(2a^2 \log n)) \geq 1/2 \; \forall n \geq m$. Since

$$n^{-a^2} \exp(-cn^{(1-a^2)}/\sqrt{\log n}) \leq n^{-2} \text{ for all sufficiently large } n$$

which is implied by the following fact

$$\exp\left(cn^{(1-a^2)}/\sqrt{\log n}\right)/n^{2-a^2} \to \infty \text{ for each } a \in (0, 1),$$

(*Proof* take logarithm of both sides and use the fact that

$$n^\alpha (\log n)^{-\beta} \to \infty \text{ if } \alpha > 0, \beta > 0.),$$

the desired result follows.   □

2.2 Another Result of Barndorff-Nielsen

Theorem 2.1.1 is a special case of

Theorem 2.2.1 *If $\{A_n\}_{n\geq 1}$ is a sequence of events such that $P(A_n) \to 0$ and $\sum P(A_n \cap A_{n+v_n}^c) < \infty$ for some sequence $\{v_n\}_{n\geq 1}$ of positive integers, then $P(A_n \text{ i.o.}) = 0$.*

Proof For every $k \geq 1$, define a sequence of integers $\{i_{k,n}\}_{n\geq 1}$ as follows:

$$i_{k,n} = \begin{cases} k & \text{if } n = 1; \\ i_{k,n-1} + v_{i_{k,n-1}} & \text{if } n \geq 2. \end{cases}$$

We have $P(A_n \cap A_{n+v_n}^c \ i.o.) = 0$. As

$$A_{i_{k,n}} \cap A_{i_{k,n+1}} = A_{i_{k,n}} \cap A_{i_{k,n}+v_{i_{k,n}}}^c,$$

we have, by Theorem 2.1.1, $P(A_{i_{k,n}} \ i.o.(n)) = 0$. So

$$P\left(\cap_{n=1}^\infty A_{i_{k,n}}\right) = 0 \text{ and hence } P\left(\cup_{k=1}^\infty \cap_{n=1}^\infty A_{i_{k,n}}\right) = 0.$$

The proof will be complete if we show that

$$\limsup A_n \subset \left(\cup_{k=1}^\infty \cap_{n=1}^\infty A_{i_{k,n}}\right) \cup [A_n \cap A_{n+v_n}^c \ i.o.].$$

To this end, let $\omega \in \limsup A_n$ and $\omega \notin \limsup(A_n \cap A_{n+v_n}^c)$. Then $\exists$ an integer $m \geq 1$ such that $\omega \notin A_n \cap A_{n+v_n}^c \ \forall n > m$. Let $p > m$ be such that $\omega \in A_p$. So $\omega \in A_{p+v_p} = A_{i_{p,2}}$ and hence

$$\omega \in A_{i_{p,2} + v_{i_{p,2}}} = A_{i_{p,3}},$$

and so on. Thus $\omega \in \overset{\infty}{\underset{n=1}{\cap}} A_{i_{p,n}}$ and so $\omega \in \overset{\infty}{\underset{k=1}{\cup}} \overset{\infty}{\underset{n=1}{\cap}} A_{i_{k,n}}$. $\qquad\square$

2.3 Results of Loève and Nash

We shall first discuss a result of Loève (1951), as stated in Nash (1954); it gives a necessary and sufficient condition for $P(\limsup A_n) = 0$.

The *necessary part* runs as follows: If $P(\limsup A_n) = 0$, then there exists an integer $m \geq 1$ such that whenever $n \geq m$,

$$P(A_n^c \cap A_{n+1}^c \cap \cdots \cap A_{k-1}^c) > 0 \quad \forall k > n \tag{2.3.1}$$

and

$$\lim_{n \to \infty} \sum_{k=n}^\infty p_{nk} = 0 \tag{2.3.2}$$

where for $k > n$

$$p_{nk} = P(A_k | A_n^c \cap A_{n+1}^c \cap \cdots \cap A_{k-1}^c), \text{ and } p_{nn} = P(A_n). \tag{2.3.3}$$

For a proof, first recall from the remark after the proof of Theorem 2.1.4 that

$$P(\limsup A_n) = 0 \Leftrightarrow \lim_{n \to \infty} \left[p_{nn} + \sum_{k=n+1}^{\infty} P(A_n^c \cap \cdots \cap A_{k-1}^c \cap A_k) \right] = 0.$$

As $P(\limsup A_n) = \lim_{n \to \infty} P\left(\bigcup_{k=n}^{\infty} A_k \right) = 0, \exists$ an integer $m \geq 1$ such that $P\left(\bigcup_{k=n}^{\infty} A_k \right) < 1 \forall n \geq m$. Then (2.3.1) holds, since

$$(A_n^c \cap \cdots \cap A_{k-1}^c) \supset \left(\bigcup_{k=n}^{\infty} A_k \right)^c.$$

Now note that if $k > n$,

$$p_{nk} \leq P(A_n^c \cap \cdots \cap A_{k-1}^c \cap A_k) / \left[1 - P\left(\bigcup_{k=n}^{\infty} A_k \right) \right]$$

so that

$$\sum_{k=n}^{\infty} p_{nk} \leq \left[P(A_n) + \sum_{k=n+1}^{\infty} P(A_n^c \cap \cdots \cap A_{k-1}^c \cap A_k) \right] / \left[1 - P\left(\bigcup_{k=n}^{\infty} A_k \right) \right]$$

and hence (2.3.2) follows by letting $n \to \infty$ in the above inequality.

The *converse* runs as follows: If $\exists$ an integer $m \geq 1$ such that (2.3.1) holds and if (2.3.2) holds, then $P(\limsup A_n) = 0$. For a proof, we need to only note that if $k > n$,

$$P(A_n^c \cap \cdots \cap A_{k-1}^c \cap A_k) \leq p_{nk}.$$

We next turn the main result of Nash (1954). We first introduce a notation. Let $A^0 = A^c$ and $A^1 = A$, and define

$$A^\epsilon = \bigcap_{n=1}^{\infty} A_n^{\epsilon_n}$$

where each ϵ_n is 0 or 1 and $\epsilon = (\epsilon_1, \epsilon_2, \ldots)$; here $\{A_n\}_{n \geq 1}$ is any given sequence of events. Let

$$H = \{\epsilon : \epsilon_i = 1 \text{ for finitely many values of } i\},$$
$$H_0 = \{\epsilon \in H : P\left(A_1^{\epsilon_1} \cap \cdots \cap A_n^{\epsilon_n} \right) > 0 \ \forall n \geq 1\}.$$

Then it is well known that H is countably infinite. Clearly,

$$P\left(\bigcup_{\epsilon\in H\setminus H_0} A^\epsilon\right) = 0. \tag{2.3.4}$$

Next, assume that $\epsilon \in H_0$. Then $P(A^\epsilon | A_1^{\epsilon_1} \cap \cdots \cap A_n^{\epsilon_n})$ is well defined for each $n \geq 1$, and

$$P(A^\epsilon) = P(A_1^{\epsilon_1}) \prod_{n=2}^{\infty} \left(1 - P\left(A_n^{1-\epsilon_n} | A_1^{\epsilon_1} \cap \cdots \cap A_{n-1}^{\epsilon_{n-1}}\right)\right).$$

Thus, using Theorem 8.52 on page 208 of Apostol (1974),

$$P(A^\epsilon) = 0 \Leftrightarrow \sum_{n=2}^{\infty} P\left(A_n^{1-\epsilon_n} | A_1^{\epsilon_1} \cap \cdots \cap A_{n-1}^{\epsilon_{n-1}}\right) = \infty$$
$$\Leftrightarrow \sum_{n=2}^{\infty} P\left(A_n | A_1^{\epsilon_1} \cap \cdots \cap A_{n-1}^{\epsilon_{n-1}}\right) = \infty$$

since $\epsilon \in H$ and hence $A_n^{1-\epsilon_n} = A_n$ for all sufficiently large n. Now observe that $\liminf A_n^c = \cup\{A^\epsilon | \epsilon \in H\}$ since

$$\omega \in \liminf A_n^c \Leftrightarrow \exists \text{ an integer } m(\omega) \geq 1 \text{ such that } \omega \notin A_n \forall n \geq m(\omega)$$
$$\Leftrightarrow \omega \in A^\epsilon \text{ for some } \epsilon \in H.$$

Thus,

$$P(\liminf A_n^c) = 0 \Leftrightarrow P\left(\bigcup_{\epsilon\in H_0} A^\epsilon\right) = 0 \quad \text{(by (2.3.4))}$$
$$\Leftrightarrow P(A^\epsilon) = 0 \quad \forall \epsilon \in H_0$$
$$\Leftrightarrow \sum P(A_n | A_1^{\epsilon_1} \cap \cdots \cap A_n^{\epsilon_n}) = \infty \forall \epsilon \in H_0.$$

We have thus proved the result of Nash, namely, $P(\limsup A_n) = 1$ iff $\sum P\left(A_n | A_1^{\epsilon_1} \cap \cdots \cap A_{n-1}^{\epsilon_{n-1}}\right) = \infty \forall \epsilon \in H_0$. A related result is given in Bruss (1980).

We next give an application of Nash's result. There are two urns each containing a red and b black balls. A ball is drawn at random from the first urn. This is repeated until a black ball is drawn. Each time a red ball is drawn from the first urn, the number of balls in the second run is doubled by putting in as many red balls as there are balls of either color in the second urn before. Once a black ball is drawn from the first urn, all further draws are made at random from the second urn with replacement after each draw, and no further change is made in the composition of the contents of the second urn. Let A_n be the event of drawing a black ball in the nth trial. We show below that $P(\limsup A_n) = 1$. Note that

$$P\left(A_n | A_1^{\epsilon_1} \cap A_2^{\epsilon_2} \cap \cdots \cap A_{n-1}^{\epsilon_{n-1}}\right) = \begin{cases} 1/2 & \text{if } n < k; \\ 2^{-k} & \text{if } n \geq k, \end{cases}$$

where $k > 1$ is such that a black ball is drawn for the first time at $(k-1)$th trial. (This is true, because the second urn will contain 2^k balls at the kth trial and thereafter, $2^k - 1$ red and 1 black balls.) Then $p'_n := \inf_{n>1} P(A_n|A_1 \cap \cdots \cap A_{n-1}) = 2^{-n} > 0$; but $\sum p'_n = 1 < \infty$, so that the hypothesis of Borel's criterion (stated in the historical remarks on page 18) does not hold. However,

$$\sum P\left(A_n|A_1^{\epsilon_1} \cap \cdots \cap A_{n-1}^{\epsilon_{n-1}}\right) = \infty \forall \epsilon,$$

and so $P(\limsup A_n) = 1$.

We conclude this section with a result of Martikainen and Petrov (1990).

Theorem 2.3.1 (Martikainen and Petrov 1990) *Let $0 < \alpha \le 1$.*

(a) The following are equivalent:

   (i) $P(A_n \ i.o.) \ge \alpha$.
   (ii) $\sum P(A_n \cap B) = \infty$ for any event B satisfying $P(B) > 1 - \alpha$.
   (iii) $P(A_n \cap B) > 0$ for infinitely many values of n for every event B satisfying $P(B) > 1 - \alpha$.

(b) The following are equivalent:

   (iv) $P(A_n \ i.o.) = \alpha$
   (v) Statement (ii) holds and for each $\epsilon > 0$, $\exists$ an event B_0 such that $P(B_0) > 1 - \alpha - \epsilon$ and $\sum P(A_n \cap B_0) < \infty$.
   (vi) Statement (iii) holds and for each $\epsilon > 0$, $\exists$ an event B_0 such that $P(B_0) > 1 - \alpha - \epsilon$ and $P(A_n \cap B_0) = 0$ for all sufficiently large n.

Proof (a)   See Petrov (1995, p. 201).

(b)   (v) $\Rightarrow$ (iv): Clearly, $P(A_n \ i.o.) \ge \alpha$ by (a). If possible, let $P(A_n \ i.o.) > \alpha$. Then $\exists$ an $\epsilon > 0$ such that $P(A_n \ i.o.) \ge \alpha + \epsilon$. By (a), $\sum P(A_n \cap B) = \infty$ for every event B satisfying $P(B) > 1 - \alpha - \epsilon$. This contradicts the second condition of (v).

(vi) $\Rightarrow$ (v): This is clear.

(iv) $\Rightarrow$ (vi): Clearly, Statement (iii) holds by (c). If possible, let the second condition of (vi) fail. Then $\exists$ an $\epsilon > 0$ such that $P(A_n \cap B) > 0$ for infinitely many value of n for every event B satisfying $P(B) > 1 - \alpha - \epsilon$. Then $P(A_n \ i.o.) \ge \alpha + \epsilon$ by (a), contradicting (iv). $\square$

References

T.M. Apostol, *Mathematical Analysis*, 2nd edn. (Addison–Wesley Publishing Company, Inc., Reading, 1974)

O.E. Barndorff-Nielsen, On the rate of growth of the partial maxima of a sequence of independent and identically distributed random variables. MS **9**, 383–394 (1961)

N. Balakrishnan, A. Stepanov, A generalization of the Borel–Cantelli lemma. Math. Sci. **35**, 61–62 (2010)

W. Feller, *An Introduction To Probability Theory And Its Applications*, vol. 1, 3rd edn. (revised) (Wiley, New York, 1968)

M. Loève, On almost sure convergence. in *Proceedings of the Second Berkeley Symposium*, University of California (1951), pp. 279–303

S.W. Nash, An extension of the Borel–Cantelli lemma. AMS **25**, 165–167 (1954)

F.T. Bruss, A counterpart of the Borel–Cantelli lemma. J. Appl. Prob. **17**, 1094–1101 (1980)

A.I. Martikainen, V.V. Petrov, On the Borel-Cantelli lemma. Zap. Nauch. Sem. Leningrad Otd. Mat. Inst. **184**, 200–207 (1990) (in Russian)

V.V. Petrov, *Limit Theorems of Probability Theory* (Oxford University Press, New York, 1995)

Chapter 3
Variants of the Second BCL

3.1 Pairwise Independence

We shall show here that the second Borel–Cantelli lemma holds for a sequence of events which are pairwise independent. Actually, weaker conditions will suffice.

Theorem 3.1.1 (Chandra (1999)) *If*

$$P(A_i \cap A_j)$$
$$\leq (c_1 P(A_i) + c_2 P(A_j)) P(A_{j-i}) + c_3 P(A_i) P(A_j) \quad for \ 1 \leq i < j,$$

where c_1, c_2 are non-negative reals and $c_3 \in \mathbb{R}$, and $\sum P(A_n) = \infty$, then $c :=$ $c_3 + 2(c_1 + c_2) \geq 1$ and $P(\limsup A_n) \geq 1/c$.

Proof Fix an integer $m \geq 1$. Let $s_n = \sum_{i=m}^{m+n} P(A_i)$ for $n \geq 1$. As $\sum_{i=m}^{\infty} P(A_i) = \infty$, $\exists \ n_0 \geq 1$ such that $s_n > 0 \ \forall n \geq n_0$. Then for $n \geq n_0$,

$$P\left(\bigcup_{i=m}^{\infty} A_i \right) \geq P\left(\bigcup_{i=m}^{m+n} A_i \right) \geq \frac{s_n^2}{s_n + c s_n^2 - c_3 \sum_{i=m}^{m+n} (P(A_i))^2}$$

by Lemma 1.4.1, p. 25. Letting $n \to \infty$, we get $P\left(\bigcup_{i=m}^{\infty} A_i \right) \geq 1/c$ since $s_n \to \infty$. So $c \geq 1$. Now letting $m \to \infty$, we get the desired result. $\square$

Special cases of Theorem 3.1.1 have been discussed by several authors. See, e.g., Erdös and Rényi (1959), Chow and Teicher (1997 Exercise 16, p. 102), Chung (2001 Exercise 11, p. 83), Lamperti (1963), and Petrov (2002).

Following theorem is now immediate.

Theorem 3.1.2 (Borel's Zero-One Law) *Assume that either (a) or (b) below holds:*

(a) $\{A_n\}_{n\geq 1}$ is pairwise NQD (a fortiori, $\{A_n\}$ is pairwise independent);
(b) $\exists \ \alpha \geq 0$ and $\beta \geq 0$ with $\alpha + \beta = 1$ such that

T. K. Chandra, *The Borel–Cantelli Lemma*, SpringerBriefs in Statistics, 63
DOI: 10.1007/978-81-322-0677-4_3, © The Author(s) 2012

$$P(A_i \cap A_j) \leq \frac{1}{2}(\alpha P(A_i) + \beta P(A_j))P(A_{j-i}) \quad \text{for } 1 \leq i < j.$$

Then $P(\limsup A_n) = 0$ or 1 according as $\sum P(A_n)$ converges or diverges.

(For an alternative proof of Theorem 3.1.2 (a), see p. 73 of Khoshnevisan (2007)).

The above result is only one of the many 0-1 laws of probability theory. However, it is **not** a consequence of any such 0-1 law, since it deals with a dependent sequence of events.

The next result is a classical one, and is a special case of Theorem 3.2.1 below. So we shall not prove it now; see Rényi (1970 p. 391). For applications, see Billingsely (1995 p. 89).

Theorem 3.1.3 (Erdös and Rényi (1959)) *If $\sum P(A_n) = \infty$ and*

$$\limsup \frac{\left(\sum_{i=1}^{n} P(A_i)\right)^2}{\sum\sum_{1 \leq i, j \leq n} P(A_i \cap A_j)} \geq 1, \tag{3.1.1}$$

then $P(\limsup A_n) = 1$.

The above theorem implies the extension of the second Borel-Cantelli lemma for **pairwise NQD events.** To see this, assume that $P(A_i \cap A_j) \leq P(A_i)P(A_j) \, \forall i \neq j$ and $\sum P(A_n) = \infty$. Then

$$\sum_{i=1}^{n}\sum_{j=1}^{n} P(A_i \cap A_j) \leq \sum_{i=1}^{n} P(A_i) + \sum_{1 \leq i \neq j \leq n} P(A_i)P(A_j)$$

$$= \sum_{i=1}^{n} P(A_i) + \left(\sum_{i=1}^{n} P(A_i)\right)^2 - \sum_{i=1}^{n}(P(A_i))^2$$

$$\leq \left(\sum_{i=1}^{n} P(A_i)\right)\left(1 + \sum_{i=1}^{n} P(A_i)\right)$$

so that 3.1.1 holds. Hence $P(\limsup A_n) = 1$.

Remark 3.1.1 Let $N_n = \sum_{i=1}^{n} I_{A_i}, n \geq 1$. Then

$$\text{Var}(N_n) = \sum_{1 \leq i \leq n}\sum_{1 \leq j \leq n} P(A_i \cap A_j) - \left(\sum_{i=1}^{n} P(A_i)\right)^2 \geq 0$$

so if $\sum_{i=1}^{n} P(A_i) > 0$, then the ratio in 3.1.1 is ≤ 1. Thus in Theorem 3.1.3, the lim sup on the right side of 3.1.1 is actually 1.

Example 3.1.1 (**Newman**) Let $\{X_n\}$ be iid Bernoulli variables with $P(X_1 = 1) = p, p > 0$. We say that a success run of length $m \geq 1$ occurs at trial n iff

$$X_n = 1, \ldots, X_{n+m-1} = 1, \ X_{n+m} = 0.$$

Let L_n be the length of the success run at trial n ($L_n = 0$ iff $X_n = 0$).

(a) Let $r_n \geq 0$ be integers. Show that $P(L_n > r_n \ i.o.) = 0$ or 1 according as the series $\sum_{n=1}^{\infty} p^{r_n}$ converges or diverges.

(b) $P(\limsup(L_n/\log n) = 1/(-\log p)) = 1, \ 0 < p < 1$.

Solution:

(a) Clearly, $P(L_n > r_n) = p^{r_n}$. So it suffices to show that $P(L_n > r_n \ i.o.) = 1$ if $\sum p^{r_n} = \infty$.

To this end, let $A_n = [L_n > r_n], n \geq 1$. Suppose that $j + r_j < k$; then the events $A_j = [X_j = 1, \ldots, X_{r_j+j} = 1]$ and $A_k = [X_k = 1, \ldots, X_{r_k+k} = 1]$ are independent so that $P(A_j \cap A_k) = P(A_j)P(A_k)$. If $j < k < r_j + j$ and $m = \max\{j + r_j, k + r_k\}$, then

$$A_j \cap A_k = [X_j = 1, \ldots, X_m = 1]$$

so that

$$P(A_j \cap A_k) = p^{m-j+1} \leq p^{k+r_k-j+1} = p^{k-j}P(A_k).$$

Observe next that if $s_n = \sum_{n=1}^{n} P(A_i)$,

$$\sum_{j=1}^{n} \sum_{k=1}^{n} P(A_j \cap A_k) \leq s_n + 2 \sum_{\substack{j+k\leq n \\ j+r_j<k}} P(A_j \cap A_k) + 2 \sum_{\substack{j<k\leq n \\ k\leq r_j+j}} P(A_j \cap A_k).$$

The second term on the right side is $\leq s_n^2$, while the third term on the right side is $\leq (1 - p)^{-1}s_n$. Hence

$$\text{the ratio in (3.1.1)} \geq 1/\{(1 + (1 - p)^{-1})s_n^{-1} + 1\} \to 1$$

so that 3.1.1 holds. By Theorem 3.1.3, $P(L_n > r_n \ i.o.) = 1$.

(b) Let $r_n \geq 0$ be a real, $n \geq 1$. Then $\sum p^{r_n}$ and $\sum p^{\lfloor r_n \rfloor}$ either both converge or both diverge. Consequently, by (a), $P(L_n > r_n \ i.o.) = 0$ or 1 according as $\sum p^{r_n}$ converges or diverges.

Put $\alpha = -\log p$. Let $\epsilon > 0$. Taking $r_n = ((1 + \epsilon)/\alpha) \log n$, and noting that $p^{r_n} = \exp(-\alpha r_n) = n^{-(1+\epsilon)}$, we get

$$P\left(L_n > \frac{1+\epsilon}{\alpha} \log n \ i.o.\right) = 0,$$

and so

$$P\left(\limsup \frac{L_n}{\log n} \le \frac{1+\epsilon}{\alpha}\right) = 1, \ \forall \epsilon > 0;$$

putting $\epsilon = 1/i$ for $i \ge 1$, we have

$$P\left(\limsup \frac{L_n}{\log n} \le 1/\alpha\right) = 1.$$

Next, taking $r_n = (1/\alpha)\log n$, we get

$$P\left(L_n > \frac{1}{\alpha}\log n \ i.o.\right) = 1;$$

thus

$$P\left(\limsup \frac{L_n}{\log n} \ge 1/\alpha\right) = 1. \tag{3.1.2}$$

Example 3.1.2 Let $\{X_n\}_{n\ge 1}$ be a sequence of strictly positive, integer valued iid random variables, and let $E_i = [X_1 + \cdots + X_k = i$ for some $k \ge 1]$, $i \ge 1$. Then, clearly,

$$P(E_i \cap E_j) = P(E_i)P(E_{j-i}), \ 1 \le i < j.$$

Let $\{m_n\}_{n\ge 1}$ be any subsequence satisfying $\sum_{n=1}^{\infty} P(E_{m_n}) = \infty$. Then $P(E_{m_n} i.o.(n)) \ge 1/2 > 0$ by Theorem 3.1.1. An application of the Hewitt-Savage zero-one law (see Billingsley (1995 p. 496)) now shows that $P(E_{m_n} i.o.(n)) = 1$.

3.2 Extended Rényi-Lamperti Lemma

Here we investigate those cases where $\sum P(A_n) = \infty$ but $P(\limsup A_n) > 0$. These constitute partial converses of the first Borel-Cantelli lemma.

The following result is a special case of a result of Kochen and Stone (1964). We give a direct proof. See, in this connection, Spitzer (1964 p. 319).

Theorem 3.2.1 (Extended Rényi-Lamperti Lemma) *Let*

$$\liminf \frac{\sum_{i=1}^{n} \sum_{j=1}^{n} P(A_i \cap A_j)}{\left(\sum_{i=1}^{n} P(A_i)\right)^2} = c,$$

and $\sum P(A_n) = \infty$. *Then* $c \ge 1$ *and* $P(\limsup A_n) \ge 1/c$.

(The Rényi-Lamperti lemma concludes that $P(\limsup A_n) \ge 2 - c$; see Billingsley (1991, p. 87).)

Proof Let $0 < \epsilon < 1$. Define

$$N = \sum_{n=1}^{\infty} I_{A_n}, \quad N_n = \sum_{i=1}^{n} I_{A_i}, \quad s_n = E(N_n) = \sum_{i=1}^{n} P(A_i), n \geq 1.$$

Let $B_n = [N_n \geq \epsilon s_n], n \geq 1$. Note that $[N = \infty] = \limsup A_n$ and

$$E(N_n^2) = \sum_{i=1}^{n} \sum_{j=1}^{n} P(A_i \cap A_j), \quad n \geq 1.$$

Also,

$$P(N \geq \epsilon s_n) \geq P(N_n \geq \epsilon s_n)$$
$$\geq (1 - \epsilon)^2 (E(N_n))^2 / E(N_n^2) \text{ by Paley-Zygmund's inequality.}$$

Since $s_n \uparrow \infty$,

$$P(N = \infty) = \lim P(N \geq \epsilon s_n) \geq (1 - \epsilon)^2 \limsup \frac{(E(N_n))^2}{E(N_n^2)} = (1 - \epsilon^2)/c.$$

Letting $\epsilon \to 0$, we see that the desired inequality is true.

That $c \geq 1$ follows from the fact that if $s_n > 0$, then the ratio in (3.1.1) is ≤ 1.

Alternative Proof. (Due to Yan (2006))

This is based on the Chung-Erdös inequality. We shall use the above notation.
Note as $E(N_n) \to \infty$, $E(N_n^2) \to \infty$. Now note that

$$\sum_{i,j=m+1}^{n} \sum P(A_i \cap A_j) \leq E(N_n^2) - E(N_m^2).$$

So by the Chung-Erdös inequality,

$$P\left(\bigcup_{k=m+1}^{\infty} A_k\right) \geq P\left(\bigcup_{k=m+1}^{n} A_k\right) \quad \forall n \geq m + 1$$
$$\geq (E(N_n) - E(N_m))^2 / (E(N_n^2) - E(N_m^2)).$$

So

$$P\left(\bigcup_{k=m+1}^{\infty} A_k\right) \geq \limsup_{n \to \infty} \frac{(E(N_n))^2}{E(N_n^2)}.$$

Letting $m \to \infty$, we get the desired inequality.   □

Note that if $\sum_{i=1}^{\infty} P(A_i) = \infty$,

$$\limsup \frac{\left(\sum_{i=1}^{n} P(A_i)\right)^2}{\sum_{i=1}^{n} \sum_{j=1}^{n} P(A_i \cap A_j)} = \limsup \frac{\sum_{1 \le i < j \le n} P(A_i)P(A_j)}{\sum_{1 \le i < j \le n} P(A_i \cap A_j)};$$

for, as $\sum_{i=1}^{n} P(A_i) \to \infty$, and

$$\left(\sum_{i=1}^{n} P(A_i)\right)^2 \le 2 \sum_{1 \le i < j \le n} P(A_i)P(A_j) + \sum_{i=1}^{n} P(A_i)$$

we have

$$\lim \frac{\sum_{i=1}^{n} P(A_i)}{\sum_{1 \le i < j \le n} P(A_i)P(A_j)} = 0$$

and

$$\lim \frac{\sum_{i=1}^{n} (P(A_i))^2}{\sum_{1 \le i < j \le n} P(A_i)P(A_j)} = 0.$$

Theorem 3.1.1 is a special case of Theorem 3.2.1, since then

$$\sum_{i=1}^{n} \sum_{j=1}^{n} P(A_i \cap A_j) \le s_n + c s_n^2 - c_3 \sum_{i=1}^{n} (P(A_i))^2$$

(see the proof, on p. 25, of Lemma 1.4.1) where $c = c_3 + 2(c_1 + c_2)$, and the ratio in (3.1.1) $\ge s_n^2/(s_n + c s_n^2 - c_3 \sum_{i=1}^{n} (P(A_i))^2) \to 1/c$.

Theorem 3.2.2 (Petrov (2004)) *Let* $\sum_{i=1}^{\infty} P(A_i) = \infty$ *and* H *be a real number. Put*

$$\alpha_H = \liminf \frac{\sum_{1 \le i < j \le n} (P(A_i \cap A_j) - H P(A_i)P(A_j))}{\left(\sum_{i=1}^{n} P(A_i)\right)^2}. \tag{3.2.1}$$

Then $P(\limsup A_n) \ge 1/(H + 2\alpha_H)$.

Proof (Due to Yan (2006)) Note that, if $s_n = \sum_{i=1}^{n} P(A_i)$,

$$H + 2\alpha_H = \liminf \left\{ \left(\sum_{i=1}^{n} \sum_{j=1}^{n} P(A_i \cap A_j) \right) /s_n^2 - 1/s_n \right.$$

$$\left. + H \left(\sum_{i=1}^{n} (P(A_i))^2 \right) /s_n^2 \right\}$$

$$= \liminf \frac{\sum_{1}^{n} \sum_{1}^{n} P(A_i \cap A_j)}{\left(\sum_{i=1}^{n} P(A_i)\right)^2}.$$

Thus Theorem 3.2.1 completes the proof.   □

The following result, due to Ortega and Wschebor (1983), follows from Theorem 3.2.2. This paper contains an application as well.

Theorem 3.2.3 *Let* $\sum P(A_n) = \infty$ *and* $\alpha_1 \leq 0$ *where* α_H *is given by* (3.2.1). *Then* $P(A_n \ i.o.) = 1$.

On the other hand, Theorem 3.2.2 follows from the next one.

Theorem 3.2.4 Chandra (2008) *Let* $\sum P(A_n) = \infty$ *and let*

$$\liminf \frac{\sum_{1 \leq i < j \leq n}(P(A_i \cap A_j) - a_{ij})}{\left(\sum_{1 \leq i \leq n} P(A_i)\right)^2} = L,$$

where

$$a_{ij} = (c_1 \ P(A_i) + c_2 \ P(A_j))P(A_{j-i}) + c_3 \ P(A_i)P(A_j), \quad 1 \leq i < j,$$

$c_1 \geq 0, c_2 \geq 0$ *and* c_3 *being constants* (L *may depend on* c_1, c_2 *and* c_3). *Assume that* L *is finite. Then* $c + 2L \geq 1$ *and*

$$P(A_n \ i.o.) \geq 1/(c + 2L)$$

where $c = c_3 + 2(c_1 + c_2)$.

Proof As seen in proof of Lemma 1.4.1 on p. 25,

$$\sum_{1 \leq i < j \leq n} a_{ij} \leq \frac{1}{2} \, cs_n^2 - c_3 \sum_{i=1}^{n}(P(A_i))^2,$$

where $s_n = \sum_{i=1}^{n} P(A_i), n \geq 1$. So,

$$c + 2L = \liminf \left\{ \sum_{i=1}^{n} \sum_{j=1}^{n} P(A_i \cap A_j)/s_n^2 \right.$$

$$\left. -1/s_n + c_3 \left(\sum_{i=1}^{n}(P(A_i))^2 \right)/s_n^2 \right\}$$

$$= \liminf \frac{\sum_{i=1}^{n} \sum_{j=1}^{n} P(A_i \cap A_j)}{\left(\sum_{i=1}^{n} P(A_i)\right)^2}.$$

Thus Theorem 3.2.1 completes the proof.   □

It is needless to remark here that Theorem 3.2.2 (and hence Theorem 3.2.4) implies Theorem 3.2.1 (take $H = 0$). Also, Theorem 3.2.4 implies Theorem 3.1.1.

3.3 Results of Kochen and Stone

In this section, we shall discuss the finding of Kochen and Stone (1964). We begin with an inequality based on Paley-Zygmund's inequality, p. 24.

Lemma 3.3.1 *Let each of the X_n has non-zero finite mean. Then*

$$P(\lim \sup(X_n/E(X_n)) > 0) \geq \lim \sup((E(X_n))^2/E(X_n^2)).$$

Proof Put $Y_n = X_n/E(X_n)$, $n \geq 1$. Then $E(Y_n) = 1$. Let $0 < c < 1$. Then

$$P(\lim \sup Y_n \geq c) \geq \lim \sup P(Y_n \geq c)$$
$$\geq (1 - c)^2 \lim \sup(E(Y_n^2))^{-1} \quad \text{by Theorem 1.4.3 (b)}.$$

Now let $c = 1/m$ and then let $m \to \infty$.   □

Now assume that each of the X_n has non-zero mean and positive finite second moment, and that

$$\lim \sup((E(X_n))^2/E(X_n^2)) \geq 1. \tag{3.3.1}$$

Then $\lim \inf (\text{var}(X_n)/(E(X_n))^2) \leq 0$. We now show that $\exists$ a subsequence $\{n_k\}_{k \geq 1}$ of positive integers such that $\forall k \geq 1$

$$\text{var}(X_{n_k})/(E(X_{n_k}))^2 \leq 1/k^2. \tag{3.3.2}$$

To this end, we shall use the mathematical induction on k. As

$$\lim \inf(\text{var}(X_n)/(E(X_n))^2) < 1,$$

we must have $\text{var}(X_n)/(E(X_n))^2 \leq 1$ for infinitely many values of n. So $\exists$ an integer $n_1 \geq 1$ such that (3.3.2) holds for $k = 1$. Now, suppose that $\exists n_1 < n_2 < \cdots < n_m$ such that (3.3.2) holds for $k = 1, \ldots, m$. As

$$\lim \inf(\text{var}(X_n)/(E(X_n))^2) < 1/(m + 1)^2,$$

we can conclude as above that $\exists$ an integer $n_{m+1} > n_m$ such that (3.3.2) holds for $k = m + 1$.

Then $\sum \text{var}(X_{n_k})/(E(X_{n_k}))^2 < \infty$ so that

$$X_{n_k}/E(X_{n_k}) \to 1 \text{ a.s. by Theorem 1.5.1 (c), p. 26}.$$

Hence

$$\lim \inf(X_n/E(X_n)) \leq 1 \leq \lim \sup(X_n/E(X_n)) \text{ a.s.}$$

If, furthermore, $E(X_n) \to \infty$, then it follows that $\lim \sup X_n = \infty$ a.s. We have, therefore, obtained the following **extension of the Erdös-Rényi Theorem**: Assume the setup of Theorem 3.1.3; then

$$\lim \sup(N_n/s_n) \geq 1 \; a.s.$$

where we have used the notation of the proof of Theorem 3.2.1.

Assume next the setup of Theorem 3.2.1. Using the notation of the last paragraph, note that as $E(N_n) \to \infty$,

$$\left[\lim \sup(N_n/E(N_n)) > 0\right] \subset [N_n \to \infty] = \lim \sup A_n;$$

so Lemma 3.3.1 implies Theorem 3.2.1.

Theorem 3.3.1 *Let each of the X_n have non-zero mean and positive finite second moment. Assume that $\lim \sup((E(X_n))^2/E(X_n^2)) > 0$. Then*

(a) $P(\lim \inf(X_n/E(X_n)) \leq 1) > 0$; *and*
(b) $P(\lim \sup(X_n/E(X_n)) \geq 1) > 0$.

If, in addition, $\lim \inf(X_n/E(X_n))$ and $\lim \sup(X_n/E(X_n))$ are a.s. constants, then

$$\lim \inf(X_n/E(X_n)) \leq 1 \; a.s., \text{ and } \lim \sup(X_n/E(X_n)) \geq 1 \; a.s.$$

Proof Let $Y_n = X_n/E(X_n), n \geq 1$. Then $E(Y_n) = 1$ and $M := \lim \inf E(Y_n^2) < \infty$. Let $a > 0$, and $Z_n = Y_n I_{[Y_n \leq a]}, n \geq 1$. Then

$$1 = E(Y_n) \leq E(Z_n) + E(Y_n^2 I_{[Y_n \geq a]})/a \leq E(Z_n) + E(Y_n^2)/a.$$

Thus

$$1 - M/a \leq \lim \sup E(Z_n) \leq E(\lim \sup Z_n) \leq E(\lim \sup Y_n)$$

by Fatou's lemma applied to $\{(a - Z_n)\}_{n \geq 1}$; here we have used the fact that $Z_n \leq Y_n$ which is immediate from $Y_n - Z_n = Y_n I_{[Y_n > a]} \geq a > 0$. Letting $a \to \infty$, we get $E(\lim \sup Y_n) \geq 1$ so that $P(\lim \sup Y_n \geq 1) > 0$. This establishes (b); the proof of (a) is similar. $\square$

For the next corollary, we need a definition.

Definition 3.3.1 A sequence $\{E_n\}_{n \geq 1}$ of events is called **a system of recurrent events** if there exist iid positive integer valued random variables $\{Y_n\}_{n \geq 1}$ such that for each $k \geq 1$

$$E_k = [Y_1 + \cdots + Y_j = k \text{ for some } j \geq 1].$$

For such a system, one has $P(E_i \cap E_j) = P(E_i)P(E_{j-i})$ for $1 \leq i < j$.

Corollary 1 *Let $\{E_n\}_{n \geq 1}$ be a system of recurrent events. Let $\{m_n\}_{n \geq 1}$ be a subsequence such that $\sum P(E_{m_n}) = \infty$. Then*

$$\limsup \frac{\left(\sum_{1 \le k \le n} P(E_{m_k})\right)^2}{\sum_{1 \le i < j \le n} P(E_{m_i}) P(E_{m_j - m_i})} > 0 \tag{3.3.3}$$

and so with probability 1,

$$\liminf \frac{M_n}{\sum_{1 \le k \le n} P(E_{m_k})} \le 1 \le \limsup \frac{M_n}{\sum_{1 \le k \le n} P(E_{m_k})}, \tag{3.3.4}$$

where M_n denotes the number of $E_{m_1}, \ldots, E_{m_n}$ which occur.

Proof It is immediate from the last part of Theorem 3.3.1 (take $X_n = M_n$) and the Hewitt-Savage 0-1 law (see, e.g., Billingsley (1995, p. 496)). See, also, Example 3.1.2. □

Example 3.3.1 Let E_k be the event that the simple random walk in one dimension is at the origin at time $2k$ $(k \ge 1)$. The E_k form a system of recurrent events and $P(E_k) \sim (\pi k)^{-1/2}$. If m_n is the nth prime number, then $\sum_{n \ge 1} m_n^{-1/2} = \infty$. Thus, with probability 1, the simple random walk is at the origin at time $2p$ for infinitely many primes p. The same method shows that this result holds for the simple random walk in $\mathbb{R}^2$ where $P(E_k) \sim (\pi k)^{-1}$.

Example 3.3.2 Let E_k be the event that the simple random walk in $\mathbb{R}^3$ hits the point $(k, 0, 0), k \ge 1$. Then $P(E_k) \sim c k^{-1}$ for some positive constant c (see, e.g., Itô and McKean (1960)). The E_k are not recurrent events, but one has

$$P(E_i \cap E_j) \le (P(E_i) + P(E_j)) P(E_{j-i}) \text{ for } 1 \le i < j.$$

Let m_n be the nth prime number. Then the Hewitt-Savage 0-1 law implies that (3.3.4) is valid. In particular, with probability 1, the random walk visits $(p, 0, 0)$ for an infinite number of primes p. This result was first suggested by Itô and McKean (1960) and was verified by Erdös (1961) and Mckean (1961).

3.4 Results of Chandra (2008)

In this section, we derive another version of the second Borel–Cantelli lemma under a suitable dependence condition using Chebyshev's inequality.

Lemma 3.4.1 *Let $\{X_n\}_{n \ge 1}$ be a sequence of non-negative random variables with finite $E(X_n^2)$, and put $S_n = \sum_{i=1}^n X_i, n \ge 1$. Assume that $\sum E(X_n) = \infty$ and*

$$\liminf \text{var}(S_n)/(E(S_n))^2 = 0. \tag{3.4.1}$$

Then $P\left(\sum_{n=1}^{\infty} X_n = \infty\right) = 1$.

Proof Note that

$$P\left(\sum_{n=1}^{\infty} X_n < \infty\right) = \lim P\left(\sum_{n=1}^{\infty} X_n \leq \frac{1}{2}E(S_n)\right) \text{ as } E(S_n) \uparrow \infty$$

$$\leq \lim\inf P(S_n \leq \frac{1}{2}E(S_n))$$

$$\leq \lim\inf P(|S_n - E(S_n)| \geq \frac{1}{2}E(S_n))$$

$$\leq \lim\inf\left(\frac{4\text{var}(S_n)}{(E(S_n))^2}\right) = 0. \qquad \square$$

Lemma 3.4.2 *Let $\{X_n\}_{n\geq 1}$ be a sequence of non-negative random variables with finite $E(X_n^2)$, and put $S_n = \sum_{i=1}^{n} X_i, n \geq 1$. Assume that $\sum E(X_n) = \infty$.*

(a) Assume, furthermore, that

$$\sum_{i=1}^{n} E(X_i^2) \leq k_n E(S_n) \forall n \geq 1, \tag{3.4.2}$$

$$\sum_{j=2}^{n}\sum_{i=1}^{n} \text{cov}(X_i, X_j) - \frac{1}{2}\sum_{i=1}^{n}(E(X_i))^2 \leq c_n E(S_n) \, \forall n \geq 2 \tag{3.4.3}$$

and

$$\lim\inf((k_n + 2c_n)/E(S_n)) = 0. \tag{3.4.4}$$

Then $P\left(\sum_{n=1}^{\infty} X_n = \infty\right) = 1$.
If $0 \leq X_n \leq k_n \, \forall n \geq 1$ where $\{k_n\}_{n\geq 1}$ is nondecreasing, then (3.4.2) holds.

(b) If $0 \leq X_n \leq k_n \forall n \geq 1$ where $\{k_n\}_{n\geq 1}$ is nondecreasing, and $\{q(n)\}_{n\geq 1}, \{a_n\}_{n\geq 1}$ and $\{b_n\}_{n\geq 1}$ are non-negative sequences such that

$$\text{cov}(X_i, X_j) \leq q(j-i)(a_i + b_j) + \frac{(E(X_j))^2}{2(j-1)} \text{ if } 1 \leq i < j, \tag{3.4.5}$$

$$\lim\inf\left(\frac{\sum_{i=1}^{n-1} q(i)\left(\sum_{i=1}^{n-1} a_i + \sum_{j=2}^{n} b_j\right)}{(E(S_n))^2}\right) = 0, \tag{3.4.6}$$

and

$$k_n/E(S_n) \to 0, \tag{3.4.7}$$

then $P\left(\sum_{n=1}^{\infty} X_n = \infty\right) = 1$.

Proof (a) This is immediate from Lemma 3.4.1: For, if $n \geq 2$

$$\text{var}(S_n) = \sum_{i=1}^{n} E(X_i^2) + 2\sum_{j=2}^{n}\sum_{i=1}^{j-1} \text{cov}(X_i, X_j) - \sum_{i=1}^{n}(E(X_i))^2$$
$$\leq (k_n + 2c_n)E(S_n).$$

For the last part, note that $\sum_{i=1}^{n} E(X_i^2) \leq \sum_{i=1}^{n} E(X_i k_i) \leq k_n E(S_n)$.

(b) This follows from Part (a) and the following observation: If $n \geq 2$,

$$\sum_{j=2}^{n}\sum_{i=1}^{j-1} \text{cov}(X_i, X_j) - \frac{1}{2}\sum_{i=2}^{n}(E(X_i))^2$$

$$= \sum_{j=2}^{n}\sum_{i=1}^{j-1}\left(\text{cov}(X_i, X_j) - \frac{(E(X_j))^2}{2(j-1)}\right)$$

$$\leq \sum_{j=2}^{n}\sum_{i=1}^{j-1} q(j-i)(a_i + b_j)$$

$$= \sum_{k=1}^{n-1} q(k) \sum_{j=k+1}^{n} (a_{j-k} + b_j)$$

$$\leq \sum_{k=1}^{n-1} q(k) \left(\sum_{i=1}^{n-1} a_i + \sum_{j=2}^{n} b_j\right). \qquad \square$$

The following result is now obvious.

Theorem 3.4.2 **(a)** *Let* $\{A_n\}_{n\geq 1}$ *be a sequence of events such that* $\sum P(A_n) = \infty$, *and for each* $1 \leq i < j$,

$$P(A_i \cap A_j) - P(A_i)P(A_j) - \frac{(P(A_i))^2}{2(j-1)}$$
$$\leq q(j-i)[P(A_i) + P(A_{i+1}) + P(A_j) + P(A_{j-1})], \qquad (3.4.8)$$

and

$$\liminf\left(\frac{\sum_{i=1}^{n-1} q(i)}{\sum_{i=1}^{n} P(A_i)}\right) = 0\left(\text{a fortiori, } \sum_{i=1}^{\infty} q(i) = \infty\right).$$

Then $P(\limsup A_n) = 1$.

(b) *If* $\sum P(A_n) = \infty$, *(3.4.8) holds for* $1 \leq i < j$ *and* $\exists$ *an integer* $m \geq 1$ *such that*

$$P(A_i \cap A_j) \leq P(A_i)P(A_j) \ if \ |i - j| > m$$

then $P(\limsup A_n) = 1$.

(In Lemma 3.4.2 (b), take $a_m = E(X_m) + E(X_{m+1})$, $b_m = E(X_m) + E(X_{m-1})$ where $X_n = I_{A_n}$ for $n \geq 1$, so that

$$\sum_{i=1}^{n-1} a_i \leq 2E(S_n), \sum_{j=2}^{n} b_j \leq 2E(S_n).)$$

3.5 A Weighted Version of BCL

We first note an extension of Paley-Zygmund's inequality. So let $b \leq E(X)$, $E(X)$ be finite and $P(X = 0) < 1$. Let $p > 1$. Then

$$P(X > b) \geq [(E(X) - b)^p / E(|X|^p)]^{1/(p-1)}. \tag{3.5.1}$$

For a proof, note that

$$0 \leq E(X) - b \leq E(X I_{[X>b]}) \leq E(|X| I_{[X>b]}) \leq (E(|X|^p))^{1/p} (P(X > b))^{(p-1)/p}$$

by Hölder's inequality, and we are done.

Xie (2008) stated (with a wrong proof) a bilateral inequality on the Borel–Cantelli lemma. Later Xie (2009) stated (with an incomplete proof) the following general bilateral inequality for a bounded non-negative sequence of random variables. See, also, Hu et al. (2009).

Theorem 3.5.1 *Let $\{X_n\}_{n \geq 1}$ be a uniformly bounded sequence of non-negative random variables, and assume that $\sum E(X_n) = \infty$. Then for $p > 1$ or $0 < p < 1$,*

$$P(\limsup\{X_n \neq 0\}) \geq \limsup_{n \to \infty} T_{n,p}; \tag{3.5.2}$$

for $p < 0$

$$P(\limsup\{X_n \neq 0\}) \leq \liminf_{n \to \infty} T_{n,p}. \tag{3.5.3}$$

Here

$$T_{n,p} = [E(S_n^p)/(E(S_n))^p]^{1/(1-p)} \tag{3.5.4}$$

where $S_n = \sum_{i=1}^{n} X_i$, $n \geq 1$.

However, $T_{n,p} \geq 1$ if $p < 0$ by (1.4.5), p. 23. Hence the upper bound in (3.5.3) is trivial. (Note that $T_{n,p} \leq 1$ for $p > 0$ and $p \neq 1$.)

We now prove an extension of the above theorem when $p > 1$.

Theorem 3.5.2 (Liu (2011)) *Let $\{X_n\}_{n \geq 1}$ be a sequence of non-negative random variables satisfying $\sum E(X_n) = \infty$. Then for $p > 1$,*

$$P\left(\sum_{n=1}^{\infty} X_n = \infty\right) \geq \limsup_{n\to\infty} T_{n,p}$$

where $T_{n,p}$ is given by (3.5.4).

Proof Let $S_n = \sum_{i=1}^{n} X_i$, $n \geq 1$, and $0 < a < 1$. Then

$$P\left(\sum_{n=1}^{\infty} X_n = \infty\right) = \lim_{n\to\infty} P\left(\sum_{n=1}^{\infty} X_n > aE(S_n)\right) \text{ as } E(S_n) \uparrow \infty$$

$$\geq \limsup P(S_n > aE(S_n))$$

$$\geq (1-a)^{p/(p-1)} \limsup_{n\to\infty} T_{n,p}$$

by (3.5.1) with $X = S_n$ and $b = aE(S_n)$. As $a \in (0,1)$ is arbitrary, the proof is complete. $\qquad\square$

Let X_n be as in Theorem 3.5.2. As

$$\left[\sum_{n=1}^{\infty} X_n = \infty\right] \subset \limsup[X_n \neq 0],$$

we have, taking $p = 2$ in Theorem 3.5.2,

$$P(\limsup[X_n \neq 0])$$

$$\geq \limsup \left(\frac{\sum_{i,j=1}^{n} w_i w_j E(X_i)E(X_j)}{\sum_{i,j=1}^{n} w_i w_j E(X_i X_j)}\right)$$

where each w_n is **nonnegative and nonrandom**. The next result generalizes the above inequality, and is based on ideas of Feng, Li, and Shen (2009).

Theorem 3.5.3 (Weighted version of the extended Rényi-Lamperti Lemma) *Let* $\{X_n\}_{n\geq 1}$ *be a sequence of integrable random variables such that*

$$\lim_{n\to\infty} \sum_{i=1}^{n} X_i = \infty \text{ or } -\infty. \tag{3.5.5}$$

Then

$$P(\limsup[X_n \neq 0]) \geq \limsup \left(\frac{\left(\sum_{i=1}^{n} E(X_i)\right)^2}{\sum_{i,j=1}^{n} E(X_i X_j)}\right). \tag{3.5.6}$$

Proof We shall first prove the following facts.

(i) If the matrix,

$$\begin{bmatrix} A & C \\ C^T & B \end{bmatrix}$$

is positive semi-definite where A and B are square matrices, then

$$(\Gamma(C))^2 \le \Gamma(A)\Gamma(B),$$

where, for any matrix E, $\Gamma(E)$ is the sum of all its entries.

(ii) Under (3.5.5),

$$\lim_{n \to \infty} \left(\frac{\sum_{i,j=1}^n E(X_i X_j)}{\sum_{i,j=2}^n E(X_i X_j)} \right) = 1; \text{ and} \qquad (3.5.7)$$

$$\limsup \frac{\left(\sum_{i=1}^n E(X_i)\right)^2}{\sum_{i,j=1}^n E(X_i X_j)} = \limsup \frac{\left(\sum_{i=m}^n E(X_i)\right)^2}{\sum_{i,j=m}^n E(X_i X_j)} \ \forall \ m \ge 2.$$

To prove (i), note that for reals x, y,

$$0 \le (x, \ldots, x, y, \ldots, y) \begin{pmatrix} A & C \\ C^T & B \end{pmatrix} (x, \ldots, x, y, \ldots, y)^T$$
$$= \Gamma(A)x^2 + 2\Gamma(C)xy + \Gamma(B)y^2.$$

To prove (ii), note that the matrix

$$F_n := \left(E(X_i X_j) \right)_{n \times n} = \begin{bmatrix} A_{1 \times 1} & C_n \\ C_n^T & B_{(n-1) \times (n-1)} \end{bmatrix}$$

is positive semi-definite. So by (i), $(\Gamma(C_n))^2 \le A\Gamma(B_n) \ \forall \ n \ge 1$.

By Schwarz's inequality, $\Gamma(B_n) \ge \left(\sum_{i=2}^n E(X_i)\right)^2 \to \infty$, and so $A/\Gamma(B_n) \to 0$ and $\Gamma(C_n)/\Gamma(B_n) \to 0$. So

$$\frac{\Gamma(F_n)}{\Gamma(B_n)} = \frac{A + \Gamma(B_n) + 2\Gamma(C_n)}{\Gamma(B_n)} \to 1,$$

establishing (3.5.7). The last part now follows.

We now prove the theorem. Note that

$$P(\limsup[X_n \ne 0])$$

$$= \lim_{m \to \infty} \lim_{n \to \infty} P\left(\bigcup_{k=m}^n [X_k \ne 0] \right)$$

$$\ge \lim_{m \to \infty} \limsup_{n \to \infty} \left(\frac{\left(\sum_{k=m}^n E(X_k)\right)^2}{\sum_{i,j=m}^n E(X_i X_j)} \right)$$

$$= \limsup_{n \to \infty} \left(\frac{\left(\sum_{k=1}^{n} E(X_k) \right)^2}{\sum_{i,j=1}^{n} E(X_i X_j)} \right). \qquad \square$$

Remark 3.5.1 Replacing each X_i by $w_i X_i$ for $i \geq 1$ where $\{w_n\}_{n \geq 1}$ is a sequence of **real-valued random weights**, one gets the weighted version of the extended Rényi-Lamperti lemma. This inequality is sharp, as shown by the following example.

Let A, B be two events such that $P(A \cup B) > 0$. Let

$$A_{3n-2} = A, \quad A_{3n-1} = A_{3n} = B \quad \text{for } n \geq 1.$$

Take $X_n = I_{A_n} \; \forall \, n \geq 1$ and consider the weight sequence $\{w_n\}_{n \geq 1}$ as

$$1, 1, -I_A, 1, 1, -I_A, 1, 1, -I_A, \ldots$$

Then $\sum_{n=1}^{\infty} E(w_n X_n) = \lim_{n \to \infty} n \, P(A \cup B) = \infty$, and $\limsup A_n = A \cup B$. By the above-mentioned inequality,

$$P(\limsup A_n) \geq \lim_{n \to \infty} \frac{n^2 (P(A) + P(B) - P(A \cap B))^2}{n^2 (P(A) + P(B) - P(A \cap B))}$$
$$= P(A \cup B) = P(\limsup A_n).$$

3.6 Weakly $*$-Mixing Sequence

This section is based on Blum et al. (1963).

Definition 3.6.1 A sequence $\{X_n\}_{n \geq 1}$ of random variables is called **weakly $*$-mixing** if $\exists \, \delta > 0$ and integers $N \geq 1, k \geq 1$ such that

$$P(A \cap B) \geq \delta P(A) P(B), \quad \forall A \in \sigma(X_N, \ldots, X_n), B \in \sigma(X_{n+k}), \; \forall n \geq N.$$

Theorem 3.6.1 *Let $\{A_n\}_{n \geq 1}$ be a sequence of events such that $\sum P(A_n) = \infty$ and $\{I_{A_n}\}_{n \geq 1}$ is weakly $*$-mixing. Then $P(\limsup A_n) = 1$.*

Proof Let δ, N, k be as in Definition 3.6.1. We can assume that $\delta < 1$. Get $j \in \{1, \ldots, k\}$ such that $\sum P(A_{nk+j}) = \infty$; this is possible since $\sum P(A_n) = \infty$. Put $D_n = A_{nk+j}, n \geq 1$. It suffices to show that $P(\limsup D_n) = 1$. If not, $\exists$ an integer m such that $mk \geq N$ and $P(\cup_{i=m}^{\infty} D_i) < 1$. Fix an integer $t \geq 2$; then

$$P\left(\cup_{i=m}^{\infty} D_i\right) \geq P\left(\cup_{i=m}^{m+t} D_i\right)$$
$$= P(D_m) + P(D_m^c \cap D_{m+1}) + P(D_m^c \cap D_{m+1}^c \cap D_{m+2})$$
$$+ \cdots + P(D_m^c \cap D_{m+1}^c \cap \cdots \cap D_{m+t-1}^c \cap D_{m+t})$$
$$\geq \delta \, P(D_m) + \delta P(D_m^c) P(D_{m+1}) + \delta P(D_m^c \cap D_{m+1}^c) P(D_{m+2})$$
$$+ \cdots + \delta P(D_m^c \cap \cdots \cap D_{m+t-1}^c) P(D_{m+t})$$
$$\geq \delta P(D_m^c \cap \cdots \cap D_{m+t-1}^c) \sum_{i=m}^{m+t} P(D_i)$$
$$\geq \delta \left(1 - P(\cup_{i=m}^{\infty} D_i)\right) \sum_{i=m}^{m+t} P(D_i).$$

Letting $t \to \infty$, we get

$$P\left(\bigcup_{i=m}^{\infty} D_i\right) \geq \delta \left(1 - P(\bigcup_{i=m}^{\infty} D_i)\right) \sum_{i=m}^{\infty} P(D_i) = \infty$$

as $P\left(\cup_{i=m}^{\infty} D_i\right) < 1$ and $\sum_{i=m}^{\infty} P(D_i) = \infty$. This is a contradiction. $\qquad \square$

Remark 3.6.2 Examples 1.6.4, 1.6.15, and 1.6.16 hold if the assumption '$\{X_n\}_{n \geq 1}$ are pairwise independent' is replaced by '$\{X_n\}_{n \geq 1}$ are weakly *-mixing'.

Some results (with applications) related to Theorem 3.6.1 are obtained by Cohn (1972) and Yoshihara (1979).

3.7 Results of Fischler

We first state a definition. Let $(\Omega, \mathcal{A}, P)$ be a probability space.

Definition 3.7.1 A sequence of events $\{B_n\}_{n \geq 1}$ is said to be *mixing of density* α, if $P(B_n \cap A) \to \alpha P(A) \, \forall A \in \mathcal{A}$.

The above definition is due to Rényi (1958). Note that this implies $P(B_n) \to \alpha$ (and so $0 \leq \alpha \leq 1$).

Lemma 3.7.1 *If $\{B_n\}_{n \geq 1}$ is mixing of density α, and $P(\limsup B_n) < 1$, then $\exists$ an event D such that $P(D) > 0$ and $P(B_n \cap D) = 0$ for all sufficiently large n.*

Proof Let $E = \liminf B_n^c$. Then $P(E) > 0$. For each finite subset S of $\mathbb{N}$, the set of all natural numbers, let

$$D_S = \left(\bigcap_{i \in S} B_i\right) \cap \left(\bigcap_{i \notin S} B_i^c\right).$$

Then it follows that $S \neq S' \Rightarrow D_S \cap D_{S'} = \phi$, and that $\cup D_S = E$. As the set of all such S's is countable, we must have $P(D_{S_0}) > 0$ for some such S_0. Put $D = D_{S_0}$. This D works.   □

Theorem 3.7.1 *Let $\{B_n\}_{n \geq 1}$ be mixing with density α.*

(a) If $\alpha > 0$, then $P(\limsup B_n) = 1$.
(b) If $\alpha < 1$, then $P(\liminf B_n) = 0$.

Proof (a) Suppose that $P(\limsup B_n) < 1$. This will lead to a contradiction. By the above lemma, $\exists$ an event D such that $P(D) > 0$ and $P(B_n \cap D) \to 0$. But $P(B_n \cap D) \to \alpha P(D)$ so that $\alpha P(D) = 0$. This is a contradiction.
(b) Note that $\{B_n^c\}$ is mixing with density $(1 - \alpha)$.   □

Definition 3.7.2 A sequence of events $\{B_n\}_{n \geq 1}$ is called stable with local density α if $P(B_n \cap A) \to \int_A \alpha dP \; \forall A \in \mathcal{A}$.

The above definition is due to Rényi (1963). The following is an analog of Theorem 3.7.1 for stable sequences.

Theorem 3.7.2 *If $\{B_n\}_{n \geq 1}$ is stable with local density α, then*

$$P(\limsup B_n) \geq P(\alpha > 0), \quad P(\liminf B_n) \leq 1 - P(\alpha < 1).$$

Proof To establish the first inequality, note that we can assume $P(\alpha > 0) > 0$. Suppose that $P(\limsup B_n) < P(\alpha > 0) \leq 1$. By the above lemma, $\exists$ an event D such that $P(D) > 0$ and $P(B_n \cap D) \to 0$. So $\int_D \alpha dP = 0$. As $P(D) > 0$ and $P(\alpha > 0) > 0$, we must have $\int_D \alpha dP > 0$. This is a contradiction. The second inequality now follows, since $\{B_n^c\}_{n \geq 1}$ is stable with local density $(1 - \alpha)$.   □

Definition 3.7.3 A sequence of events $\{B_n\}_{n \geq 1}$ is called mixing if

$$P(B_n \cap A) - P(B_n)P(A) \to 0 \; \forall A \in \mathcal{A}.$$

Clearly, mixing with density α implies mixing. Also, if $P(B_n) \to 0$ then $\{B_n\}_{n \geq 1}$ is mixing.

Theorem 3.7.3 *Let $\{B_n\}_{n \geq 1}$ be mixing and $\sum P(B_n) = \infty$.*

(a) If $P(B_n) \not\to 0$, then $P(\limsup B_n) = 1$.
(b) If $P(B_n) \to 0$, then $P(\limsup B_n)$ can be arbitrary.

Proof (a) Note that $\liminf P(B_n) < \limsup P(B_n)$ so that at least one of these two numbers is not 0. Consequently, $\exists$ a subsequence $\{B_{n_k}\}_{k \geq 1}$ such that $P(B_{n_k}) \to \alpha$ for some $\alpha > 0$. But then $\{B_{n_k}\}_{k \geq 1}$ will be mixing with density α. Theorem 3.7.1 (a) now completes the proof.
(b) Consider the unit interval [0,1] with the Lebesgue measure. Let $0 \leq t \leq 1$. Split [0,t] into two equal parts B_1, B_2; then split $[0, t]$ into four equal parts

B_3, B_4, B_5, B_6, and so on. Then $\sum P(B_n) = \infty$, and $\{B_n\}_{n\geq 1}$ is mixing (see, also, the remark below). But $\limsup B_n = [0, t]$.   □

Remark 3.7.1 If the defining condition of mixing holds for each $A = B_k, k = 1, 2, \ldots$, then $\{B_n\}_{n\geq 1}$ is mixing; see Rényi (1958). It follows that **pairwise independence implies mixing**.

The above results are due to Fischler (1967b). For an alternative proof of Theorem 3.7.1, see Fischler (1967a).

3.8 Results of Martikainen and Petrov

Following Martikainen and Petrov (1990), we first discuss properties of a characteristic, $d(X)$, of a random variable X. Let

$$d(X) = (E(X))^2/E(X^2),$$

provided $P(X = 0) < 1$ and $E(|X|) < \infty$. Then $0 \leq d(X) \leq 1$ and $d(X)$ depends on X only through the distribution of X. We have, furthermore,

(a) $d(X + Y) \geq \min(d(X), d(Y))$ if $X \geq 0, Y \geq 0$;
(b) $d(X_1 + \cdots + X_n) \geq \left(\sum_{i=1}^{n} E(X_i)\right) / \left(1 + \sum_{i=1}^{n} E(X_i)\right)$, provided
    cov $(X_i, X_j) \leq 0 \ \forall i \neq j$ and $E(X_i^2) \leq E(X_i) \ \forall i$.

For a proof, note that (a) follows from

$$d(X + Y) \geq (E(X) + E(Y))^2/(E(X^2) + E(Y^2) + 2(E(X^2)E(Y^2))^{1/2})$$
$$= (E(X) + E(Y))^2/(\sqrt{E(X^2)} + \sqrt{E(Y^2)})^2$$
$$\geq (\min(E(X)/\sqrt{E(X^2)}, E(Y)/\sqrt{E(Y^2)}))^2$$

since if $E(X)/\sqrt{E(X^2)} \leq E(Y)/\sqrt{E(Y^2)}$, then

$$(\sqrt{E(Y^2)} + \sqrt{E(X^2)})/\sqrt{E(X^2)} \leq (E(Y) + E(X))/E(X),$$

while (b) follows from

$$E(X_1 + \cdots + X_n)^2 = \sum_{i \neq j} E(X_i X_j) + \sum_{i=1}^{n} E(X_i^2)$$
$$\leq \sum_{i \neq j} E(X_i)E(X_j) + \sum_{i=1}^{n} E(X_i)$$

$$\leq \left(\sum_{i=1}^{n} E(X_i) \right)^2 + \sum_{i=1}^{n} E(X_i).$$

Anděl and Dupač (1989) have obtained the following result: For a given sequence of events $\{A_n\}_{n\geq 1}$, assume that there exists a sequence of independent events $\{B_n\}_{n\geq 1}$ such that $A_n \subset B_n \ \forall n \geq 1$, $P(A_n)/P(B_n) \to 1$, and $\sum P(B_n) = \infty$; then $P(A_n \ i.o.) = 1$. Corollary 1 below extends it.

Theorem 3.8.1 (Martikainen and Petrov (1990)) *Let $\{A_n\}_{n\geq 1}$ be a sequence of events. Assume that exists a sequence of events $\{B_n\}_{n\geq 1}$ such that $P(B_n \cap A_n^c)/P(B_n) \to 0$ and $\sum P(B_n) = \infty$. Then $P(A_n \ i.o.) \geq d_M$ for each $M > 0$ where*

$$d_M = \lim_{k\to\infty} \sup d \left(\sum_{j\in S} I_{B_j} \right)$$

the supremum being taken over all finite subsets S of $\{k, k+1, k+2, \ldots\}$ such that $\sum_{j\in S} P(B_j) \leq M$.

Proof Let $k \geq 1$ be an integer. Then $\exists$ a finite subset S_k of $\{k, k+1, \ldots\}$ such that $\sum_{j\in S_k} P(B_j) \leq M$ and $d \left(\sum_{j\in S_k} I_{B_j} \right) \geq \sup d \left(\sum_{j\in S} I_{B_j} \right) - k^{-1}$. Now

$$P \left(\bigcup_{j\in S_k} B_j \right) \geq d \left(\sum_{j\in S_k} I_{B_j} \right) \to d_M \text{ as } k \to \infty.$$

But $\sum_{j\in S_k} P(B_j \cap A_j^c) \to 0$ as $k \to \infty$, and

$$P \left(\bigcup_{j\in S_k} A_j \right) \geq P \left(\bigcup_{j\in S_k} B_j \right) - \sum_{j\in S_k} P(B_j \cap A_j^c),$$

$$\text{as} B_j \subset A_j \cup (B_j \cap A_j^c) \ \forall j.$$

Therefore,

$$P(A_n \ i.o.) \geq \lim \sup P \left(\cup_{j\in S_k} A_j \right) \geq d_M. \qquad \square$$

Corollary 1 *Let $\{A_n\}_{n\geq 1}$ be a sequence of events. Assume that $\exists$ a sequence of events $\{B_n\}_{n\geq 1}$ such that $P(B_n \cap A_n^c) = o(P(B_n))$, $\sum P(B_n) = \infty$ and $\{B_n\}_{n\geq 1}$ is pairwise NQD. Then $P(A_n \ i.o.) = 1$.*

Proof By the property (b) above, we have $d_M \geq (1 + M^{-1})^{-1} \ \forall M > 0$. So $P(A_n \ i.o.) \geq (1 + M^{-1})^{-1} \ \forall \ M > 0$. Letting $M \to \infty$, we get the desired result. $\qquad \square$

Remark 3.8.1 From the above theorem, we get the result that if $\sum P(A_n) = \infty$, then $P(A_n \; i.o.) \geq \lim_{k \to \infty} \left(\sup d \left(\sum_{j \in S} I_{A_j} \right) \right) \; \forall \; M > 0$ where the supremum is taken over all finite subsets S of $\{k, k+1, \ldots\}$ such that $\sum_{j \in S} P(A_j) \leq M$. Martikainen and Petrov (1990) remark that this extends Theorem 3.2.1. They also remark that $\exists$ a sequence of events $\{A_n\}_{n \geq 1}$ such that $\sum P(A_n) = \infty$ and

$$\limsup_{,} \frac{\left(\sum_{i=1}^n P(A_i) \right)^2}{\sum_{i=1}^n \sum_{j=1}^n P(A_i \cap A_j)} = 0,$$

but

$$\sup_{M>0} \lim_{k \to \infty} \left(\sup d \left(\sum_{j \in S} I_{A_j} \right) \right) = 1,$$

the innermost supremum being taken as above. However, they did not supply any justification.

We conclude this section with four references. In Móri and Székely (1983), one gets several lower bounds for $P(\limsup A_n)$. A similar remark holds for Amghibech (2006) as well. These papers are related to Theorem 3.2.1. A version of the Borel–Cantelli lemmas for capacities is proved in Song (2010); this paper is related to Petrov (2004). A quantitative form of the BCL has been obtained by Phillipp (1967).

References to Stirling's numbers are made in Amghibech (2006); see, in this connection, Van Lint and Wilson (2001).

References

S. Amghibech, On the Borel-Cantelli lemma and moments. Comment. Math. Univ. Carolin. **47**, 669–679 (2006)

J. Andel, V. Dupač, An extension of the Borel lemma. Comment. Math. Univ. Carolin. **30**, 403–404 (1989)

P. Billingsley, Probability and Measure, 3rd edn. (Wiley, New York, 1995), Second Edition 1991. First Edition 1986

J.R. Blum, D.L. Hanson, L.H. Koopmans, On the strong law of large numbers for a class of stochastic processes. ZWVG **2**, 1–11 (1963)

T.K. Chandra, *A First Course in Asymptotic Theory of Statistics* (Narosa Publishing House Pvt. Ltd., New Delhi, 1999)

T.K. Chandra, Borel-Cantelli lemma under dependence conditions. SPL **78**, 390–395 (2008)

Y.S. Chow, H. Teicher, *Probability Theory*, 3rd edn. (Springer, New York, 1997)

K.L. Chung, *A Course in Probability Theory*, 3rd edn. (Academic Press, New York, 2001)

H. Cohn, On the Borel-Cantelli lemma. IJM **12**, 11–16 (1972)

P. Erdös, A problem about prime numbers and the random walk II. IJM **5**, 352–353 (1961)

P. Erdös, A. Rényi, On Cantor's series with convergent $\sum 1/q_n$. Ann. Univ. Sci. Budapest Eötvös. Sect. Math. **2**, 93–09 (1959)

C. Feng, L. Li, J. Shen, On the Borel-Cantelli lemma and its generalization. C.R. Acad. Sci. Paris Ser. I **347**, 1313–1316 (2009)

R.M. Fischler, The strong law of large numbers for indicators of mixing sets. Acte Math. Acad. Sci. Hung. **18**, 71–81 (1967a)

R.M. Fischler, Borel-Cantelli type theorems for mixing sets. Acta Math. Acad. Sci. Hung. **18**, 67–69 (1967b)

S.H. Hu, X.J. Wong, X.Q. Li, Y.Y. Zhang, Comments on the paper: A bilateral inequality on the Borel-Cantelli lemma. SPL **79**, 889–893 (2009)

K. Itô, H.P. McKean Jr, Potentials and random walk. IJM **4**, 119–132 (1960)

D. Khoshnevisan, *Probability* (American Mathematical Society, Providence, 2007)

S. Kochen, C. Stone, A note on the Borel-Cantelli lemma. IJM **8**, 248–251 (1964)

J. Lamperti, Wiener's test and Markov chains. J. Math. Anal. Appl. **6**, 58–66 (1963)

J. Liu, A note on the bilateral inequality for random variable sequence, P.R. China, Technical Report (2011)

A.I. Martikainen, V.V. Petrov, On the Borel-Cantelli lemma. Zap. Nauch. Sem. Leningr. Otd. Mat. Inst. **184**, 200–207 (1990). (in Russian)

T.F. Móri, G.J. Székeley, On the Erdös-Rényi generalization of the Borel-Cantelli lemma. Studia Sci. Math. Hung. **18**, 173–182 (1983)

H.P. McKean Jr, A problem about prime numbers and the random walk I. IJM **5**, 351 (1961)

J. Ortega, M. Wschebor, On the sequence of partial maxima of some random sequences. Stoch. Process. Appl. **16**, 85–98 (1983)

V.V. Petrov, A note on the Borel-Cantelli lemma. SPL **58**, 283–286 (2002)

V.V. Petrov, A generalization of the Borel-Cantelli lemma. SPL **67**, 233–239 (2004)

W. Phillipp, Some metrical theorems in numnber theory. Pac. J. Math. **20**, 109–127 (1967)

A. Rényi, On mixing sequences of sets. Acta Math. Acad. Sci. Hung. **1**, 215–228 (1958)

A. Rényi, On stable sequences of events. Sankhyā Ser. A **25**, 293–302 (1963)

A. Rényi, *Probability Theory* (North-Holland Publishing Co., Amsterdam, 1970), German Edition 1962. French version 1966. New Hungarian Edition 1965

L. Song, Borel-Cantelli lemma for capacities P.R. China, Technical Report (2010)

F. Spitzer, *Principles of Random Walk* (Van Nostrand, Princeton, 1964)

J.H. van Lint, R.M. Wilson, *A Course in Combinatorics*, 2nd edn. (Cambridge University Press, Cambridge, 2001)

Y.Q. Xie, A bilateral inequality on the Borel-Cantelli lemma. SPL **78**, 2052–2057 (2008)

Y.Q. Xie, A bilateral inequality on nonnegative bounded random sequence. SPL **79**, 1577–1580 (2009)

J. Yan, A simple proof of two generalized Borel-Cantelli lemmas, in *Memorian Paul-Andre Meyer: Seminaire de Probabilitiés XXXIX*. Lecture Notes in Mathematics No. 1874. (Springer-Verlag, 2006)

K. Yoshihara, The Borel-Cantelli lemma for strong mixing sequences of events and their applications to LIL. Kodai Math. J. **2**, 148–157 (1979)

Chapter 4
A Strengthend Form of BCL

4.1 Pairwise Independence

Let $\{A_n\}_{n\geq 1}$ be a sequence of events. Put

$$N_n = \sum_{k=1}^{n} I_{A_k}, \, s_n = E(N_n) = \sum_{k=1}^{n} P(A_k), n \geq 1. \tag{4.1.1}$$

Then if $\sum_{k=1}^{\infty} P(A_k) = \infty$,

$$\left[\frac{N_n}{E(N_n)} \to 1\right] \subset [N_n \to \infty] = \limsup A_n. \tag{4.1.2}$$

To verify this, let $N_n(\omega)/E(N_n) \to 1$. Then $\exists$ an integer $m \geq 1$ such that $N_n(\omega)/E(N_n) \geq 1/2$ and $E(N_n) > 0 \, \forall n \geq m$, and hence $\forall \, n \geq m$

$$N_n(\omega) \geq \frac{1}{2} E(N_n) \to \infty$$

since $E(N_n) = \sum_{k=1}^{n} P(A_k) \to \infty$ by the given condition.

Henceforth, we shall assume that $\sum_{k=1}^{\infty} P(A_k) = \infty$, and consider additional sufficient conditions on $\{A_n\}_{n\geq 1}$ so as to guarantee

$$N_n/E(N_n) \to 1 \, a.s. \tag{4.1.3}$$

By (4.1.2), this will strengthen the conclusion of the second Borel-Cantelli lemma. The first result extends a part of Theorem 3.1.2, page 63; see, also, Corollary 1(a) of Theorem 4.1.2 below.

T. K. Chandra, *The Borel–Cantelli Lemma*, SpringerBriefs in Statistics,
DOI: 10.1007/978-81-322-0677-4_4, © The Author(s) 2012

Theorem 4.1.1 *Let* $\{A_n\}_{n\geq 1}$ *be a sequence of events such that*

(a) $\sum P(A_n) = \infty$; *and*
(b) $\{A_n\}_{n\geq 1}$ *is pairwise NQD, i.e.,*

$$P(A_i \cap A_j) \leq P(A_i)P(A_j) \, \forall \, i \neq j$$

(a fortiori, $\{A_n\}_{n\geq 1}$ *is pairwise independent). Then* (4.1.3) *holds where* N_n *is given by* (4.1.1). $\square$

The proof of the above theorem is elementary and is based on the method of subsequences which uses only Chebyshev's inequality and Theorem 1.5.1 (c), page 26. See, e.g., Durrett (2005, p. 50) and Chandra (2012, p. 118). For some applications of this method, see pp. 51–54 of Durrett (2005).

We shall show below that minor modifications of these arguments lead to a considerably general result.

Theorem 4.1.2 *Let* $\{X_n\}_{n\geq 1}$ *be a sequence of nonnegative random variables such that* $\sum_{n=1}^{\infty} E(X_n) = \infty$. *Let* $S_n = X_1 + \cdots + X_n, n \geq 1$. *Let* $\alpha_n \uparrow \infty$ *and* $r > 0$ *be a real.*

(a) If there exist constants $c > 0$ *and* $d \in [r-1, r)$ *and an integer* $p \geq 1$ *such that*

$$E|S_n - \alpha_n|^r \leq c\alpha_n^d \, \forall \, n \geq p \tag{4.1.4}$$

then $S_n/\alpha_n \rightarrow 1$ *a.s.*
(b) Let $\alpha_n = E(S_n), n \geq 1$. *If, whenever* $1 \leq i < j$,

$$\text{cov}(X_i, X_j) \leq q(j-i)(a_i + b_j) + \frac{(E(X_j))^2}{2(j-1)} \tag{4.1.5}$$

where $\{a_n\}_{n\geq 1}, \{b_n\}_{n\geq 1}$, *and* $\{q(n)\}_{n\geq 1}$ *are non-negative sequences, and for some* $\alpha \in [1, 2)$, *one has*

$$\sum_{i=1}^{n} E(X_i^2) = 0((E(S_n))^\alpha) \text{ as } n \rightarrow \infty, \tag{4.1.6}$$

$$\left(\sum_{i=1}^{n} q(i)\right)\left(\sum_{i=1}^{n-1} a_i + \sum_{j=2}^{n} b_j\right) = 0((E(S_n))^\alpha) \text{ as } n \rightarrow \infty, \tag{4.1.7}$$

then $S_n/E(S_n) \rightarrow 1$ *a.s.*

Proof (a) Let m be an integer $\geq p$ such that $\alpha_n > 0 \, \forall n \geq m$; such an m exists since $\alpha_n \rightarrow \infty$. Let $\beta > 1/(r - d) \geq 1$. Define inductively

$$n_1 = \inf\{n \geq 1 : \alpha_n > 1\}, n_k = \inf\{n > n_{k-1} : \alpha_n > k^\beta\} \quad \text{for} \quad k \geq 2.$$
$$(4.1.8)$$

Then each n_k is well defined, and $\{n_k\}_{k \geq 1}$ is a subsequence of natural numbers. Put $Y_n = S_n/\alpha_n, n \geq m$. Now

$$\sum_k E|Y_{n_k} - 1|^r \leq c \sum_k \alpha_{n_k}^{d-r}$$

$$\leq c \sum_k k^{\beta(d-r)} \text{ as } \alpha_{n_k} \geq k^\beta \text{ and } d < r$$

$$< \infty \text{ as } \beta(r - d) > 1.$$

By Theorem 1.5.1(c), page 26,

$$Y_{n_k} \to 1 \text{ a.s.} \tag{4.1.9}$$

We now show that

$$\left[Y_{n_k} \to 1 \text{ as } k \to \infty\right] \subset [\, Y_n \to 1 \text{ as } n \to \infty], \tag{4.1.10}$$

so that (4.1.9) and the definition of Y_n together imply that $S_n/\alpha_n \to 1$ a.s. To verify (4.1.10), let $Y_{n_k}(\omega) \to 1$ as $k \to \infty$. Let $\epsilon > 0$. We show that $\exists$ an integer $t \geq m$ such that for each $n \geq t$

$$1 - \epsilon \leq Y_n \leq 1 + \epsilon; \tag{4.1.11}$$

i.e., $Y_n(\omega) \to 1$ as $n \to \infty$.
Let $k_0 \geq 1$ be such that $|Y_{n_k}(\omega) - 1| \leq \epsilon/2$ and $n_{k_0} \geq m \; \forall \, k \geq k_0$. Let $k_1 \geq k_0$ be an integer such that

$$(1 - \epsilon/2)k^\beta/(k + 1)^\beta \geq 1 - \epsilon, \text{ and } (1 + \epsilon/2)(k + 1)^\beta/k^\beta \leq 1 + \epsilon \; \forall k \geq k_1.$$

Put $t = n_{k_1}$. Let $n \geq t$. If $n = n_k$ for some $k \geq 1$, then (4.1.11) holds. Let $n \notin \{n_k : k \geq 1\}$. Then $\exists \, k \geq k_1$ such that $n_k < n \leq n_{k+1} - 1$. By the given condition, $S_{n_k} \leq S_n \leq S_{n_{k+1}-1}$, and $k^\beta \leq \alpha_{n_k} \leq \alpha_n \leq \alpha_{n_{k+1}-1} \leq (k + 1)^\beta$. So $(1 - \epsilon/2)k^\beta/(k + 1)^\beta \leq (\alpha_{n_k}/\alpha_{n_{k+1}-1})Y_{n_k} \leq Y_n \leq Y_{n_{k+1}-1}(\alpha_{n_{k+1}-1}/\alpha_{n_k}) \leq (1 - \epsilon/2)(k + 1)^\beta/k^\beta$ so that (4.1.11) again holds.

(b) In view of Part (a), it suffices to show that (4.1.5)–(4.1.7) together imply (4.1.4). Let $m \geq 2$ be an integer such that $E(S_n) \geq 1 \; \forall n \geq m$. Then if $n \geq m$,

$$\text{var}(S_n) = \sum_{i=1}^{n} E(X_i^2) + 2 \sum_{j=2}^{n} \sum_{i=1}^{j-1} \left(\text{cov}(X_i, X_j) - \frac{(E(X_j))^2}{2(j-1)} \right)$$

$$\le \sum_{i=1}^{n} E(X_i^2) + 2 \sum_{j=2}^{n} \sum_{i=1}^{j-1} q(j-i)(a_i + b_j)$$

$$= \sum_{i=1}^{n} E(X_i^2) + 2 \sum_{j=2}^{n} \sum_{k=1}^{j-1} q(k)(a_{j-k} + b_j)$$

$$= \sum_{i=1}^{n} E(X_i^2) + 2 \sum_{k=1}^{n-1} q(k) \sum_{j=k+1}^{n} (a_{j-k} + b_j)$$

$$\le \sum_{i=1}^{n} E(X_i^2) + 2 \sum_{k=1}^{n-1} q(k) \left(\sum_{j=1}^{n-1} a_j + \sum_{j=2}^{n} b_j \right)$$

$$= 0((E(S_n))^\alpha). \qquad \qquad \square$$

Corollary 1 *Let $\{A_n\}_{n \ge 1}$ be a sequence of events such that $\sum P(A_n) = \infty$.*

(a) *If there exists an integer $m \ge 0$ such that $P(A_i \cap A_j) \le P(A_i)P(A_j)$ $\forall |i - j| > m$, then (4.1.3) holds where N_n is given by (4.1.1).*

(b) *More generally, assume that $\exists$ an $\alpha \in [1, 2)$ such that*

$$\sum_{j=2}^{n} \sum_{i=1}^{j-1} \left(P(A_i \cap A_j) - P(A_i)P(A_j) \right) - \frac{1}{2} \sum_{i=1}^{n} (P(A_i))^2 = 0(s_n^\alpha), \quad (4.1.12)$$

where $s_n = E(N_n) = \sum_{k=1}^{n} P(A_k), n \ge 1$. Then (4.1.3) holds.

Proof (a) We use Theorem 4.1.2 (a). First, assume $m = 0$. So put $X_n = I_{A_n}, n \ge 1$; then $N_n = X_1 + \cdots + X_n, n \ge 1$, and $\sum E(X_n) = \infty$, $\{E(X_n)\}_{n \ge 1}$ is bounded by 1 and cov $(X_i, X_j) \le 0$ for $i \ne j$. Then if $n \ge 2$,

$$\text{var} N_n = \sum_{i=1}^{n} \text{var}(X_i) + \sum_{i \ne j} \text{cov}(X_i, X_j)$$

$$\le \sum_{i=1}^{n} \text{var}(X_i) \le \sum_{i=1}^{n} E(X_i^2) = \sum_{i=1}^{n} E(X_i) = s_n$$

so that (4.1.4) holds with $p = 2, c = 1$ and $\alpha = 1$.

Now assume that $m \ge 1$. Then, with the above notation, for $n \ge m + 1$

$$\text{var}(N_n) \le s_n + \sum_{i,j=1,|i-j|\le m}^{n} \text{cov}(X_i, X_j)$$

$$\le s_n + \sum_{i,j=1,|i-j|\le m}^{n} P(A_i \cap A_j)$$

$$\le s_n + 2 \sum_{i=1}^{n} \sum_{k=1}^{m} P(A_i \cap A_{i+k})$$

$$\le s_n + 2 \sum_{i=1}^{n} \sum_{k=1}^{m} P(A_i) = (2 + 2m)s_n.$$

Thus, Theorem 4.1.2 (a) is, again, applicable.

(b) We again use Theorem 4.1.2 (a) with $X_n = I_{A_n}, n \ge 1$. It suffices to show that (4.1.4) holds. Let $m \ge 2$ be an integer such that $s_n := E(N_n) = \sum_{i=1}^{n} P(A_i) \ge 1 \forall n \ge m$. Then if $n \ge m$,

$$\text{var}(N_n)$$
$$= \sum_{i=1}^{n} P(A_i) - \sum_{i=1}^{n}(P(A_i))^2 + 2 \sum_{j=2}^{n}\sum_{i=1}^{j-1}(P(A_i \cap A_j) - P(A_i)P(A_j))$$
$$= s_n + 0(s_n^\alpha) = 0(s_n^\alpha).$$

Thus (4.1.4) holds. □

We next apply Theorem 4.1.2 (a) to improve the above Corollary 1 (a) on page 88 considerably.

Theorem 4.1.3 *Let $\{A_n\}_{n\ge1}$ be a sequence of events such that $\sum P(A_n) = \infty$. Assume that $\exists$ an integer $m \ge 1$, a non-negative sequence $\{q(n)\}_{n\ge1}$ and a constant $\alpha \in [1, 2)$ such that*

$$P(A_i \cap A_j) - P(A_i)P(A_j) - (P(A_j))^2/(2(j-1))$$
$$\le q(|i-j|)[P(A_i) + P(A_{i+1}) + P(A_j) + P(A_{j-1})] \; if \; |i-j| > m \quad (4.1.13)$$

and $\sum_{i=1}^{n-1} q(i) = 0(s_n^{\alpha-1})$ where $s_n = \sum_{i=1}^{n} P(A_i), n \ge 1$. Then (4.1.3) holds where N_n is as in (4.1.1).

Proof We apply Theorem 4.1.2 with $X_n = I_{A_n}, n \ge 1$; see also the proof of Corollary 1 (a). Put $a_i = P(A_i) + P(A_{i+1}), b_j = P(A_j) + P(A_{j-1}), i \ge 1, j \ge 2$. Then (4.1.12) reduces to (4.1.5); (4.1.6) holds, since $\sum_{i=1}^{n} E(X_i^2) = E(N_n) = 0((E(N_n))^\alpha)$. Finally, (4.1.7) holds, since $\sum_{i=1}^{n-1} a_i \le 2s_n, \sum_{j=2}^{n} b_j \le 2s_n$. □

For extensions of Theorem 4.1.2 (a), see Petrov 2008; 2009.

Example 4.1.1 Let $\{X_n\}_{n\geq 1}$ be a sequence of nonnegative random variables such that $E(X_i - \lambda_i)(X_j - \lambda_j) \leq 0 \; \forall \; i \neq j$ and $E(X_i - \lambda_i)^2 \leq c\lambda_i \; \forall i \geq 1$ where $c > 0$ and $\lambda_i \geq 0 \; \forall \; i \geq 1$. Assume that $\sum_{i=1}^{\infty} \lambda_i = \infty$. Then $S_n/a_n \to 1 \; a.s.$ where $S_n = X_1 + \cdots + X_n, a_n = \lambda_1 + \cdots + \lambda_n$ for $n \geq 1$.

Solution: We shall apply Theorem 4.1.2 (a) with $r = 2$ and $d = 1$. To this end, note that for $n \geq 1$

$$E(S_n - a_n)^2 = \sum_{i=1}^{\infty} E(X_i - \lambda_i)^2 + \sum_{i \neq j} E(X_i - \lambda_i)(X_j - \lambda_j)$$
$$\leq c \, a_n.$$

Example 4.1.2 Let $\{X_n\}_{n\geq 1}$ be a sequence of pairwise independent random variables such that X_n follows the Poisson distribution with mean λ_n for each $n \geq 1$. If $\sum_n \lambda_n = \infty$, then $S_n/a_n \to 1 \; a.s.$ where S_n and a_n are as in Example 4.1.1.

Solution: This is immediate from Example 4.1.1.

Remark 4.1.1 Assume, in Example 4.1.2, that the X_n is independent. Then the result is immediate from the SLLN of Kolmogorov. An alternative proof runs as follows: Let $n_j = [\lambda_j]$, the integer part of λ_j, and then replace $\{X_n\}$ by a sequence $\{Y_n\}_{n\geq 1}$ of independent random variables with $Y_1, \ldots, Y_{n_1}$ following the Poisson (1) distribution, Y_{n_1+1} following the Poisson $(\langle \lambda_1 \rangle)$ distribution, $Y_{n_1+2}, \ldots, Y_{n_1+n_2+1}$ following the Poisson (1) distribution, $Y_{n_1+n_2+2}$ following the Poisson $(\langle \lambda_2 \rangle)$ distribution, and so on; here $\langle x \rangle$ stands for the fractional part of x. Then the means of the Y_n are bounded by 1 and the proof of Theorem 4.1.1 goes through.

4.2 A Strong Law and the Second BCL

We shall use a modification, due to Chandra and Goswami (1992), of the SLLN of Csörgö et al. (1983).

Theorem 4.2.1 *Let $\{X_n\}_{n\geq 1}$ be a sequence of non-negative random variables and $\{f(n)\}_{n\geq 1}$ is a nondecreasing sequence of positive reals such that*

(a) $f(n) \to \infty$;
(b) $\exists$ *a double sequence* $\{\rho_{ij}\}$ *of nonnegative reals satisfying*

$$\text{var}(S_n) \leq \sum_{i=1}^{n} \sum_{j=1}^{n} \rho_{ij}, \quad n \geq 1; \; and$$

(c) $\sup_{n\geq 1} \left(\sum_{k=1}^{n} E(X_k)/f(n) \right) < \infty.$

Assume that $\exists\, q(m) \geq 0, c_j \geq 0 \,\forall m \geq 1, \,\forall j \geq 1$ such that

(d) $\rho_{j-m,j} \leq q(m)c_j$ *for* $m = 1, \ldots, j-1$ *and* $j \geq 2$;

(e) $\sum_{j=1}^{\infty} \rho_{jj}/(f(j))^2 < \infty$; *and*

(f) $\sum_{m=1}^{\infty} q(m) \sum_{j=m+1}^{\infty} c_j/(f(j))^2 < \infty$. *Then* $(S_n - E(S_n))/f(n) \to 0$ *a.s. where* $S_n = X_1 + \cdots + X_n, n \geq 1$.

Proof See pages 102–103 of Chandra (2012).   □

The following result is an extension of Remark 1 of Etemadi (1983).

Theorem 4.2.2 (Chandra and Ghosal (1993)) *Let* $\{A_n\}_{n\geq 1}$ *be a sequence of events satisfying*

$$P(A_i \cap A_j) - P(A_i)P(A_j) \leq q(j-i)P(A_j) \,\forall i < j \qquad (4.2.1)$$

where $q(n) \geq 0 \,\forall n \geq 1$ *and* $\sum_{n=1}^{\infty} q(n)/s_n < \infty$, s_n *being as in (4.1.1). If* $\sum P(A_n) = \infty$, *then (4.1.3) holds.*

Proof There exists an integer $m \geq 1$ such that $s_n > 0 \,\forall n \geq m$. We now apply Theorem 4.2.1 with $f(n) = s_n \uparrow \infty$. Put $X_n = I_{A_n}, n \geq 1$. Let

$$\rho_{ij} = 2(\mathrm{cov}(X_i, X_j))^+ \quad \text{if } i \leq j; \; = 0 \text{ otherwise.}$$

Then $S_n = N_n$, and Conditions (b) and (c) of Theorem 4.2.1 hold. Also,

$$\rho_{ij} \leq 2q(j-i)P(A_j) \,\forall i < j$$

so that Condition (d) holds with $c_j = 2P(A_j), j \geq 2$. Now

$$\sum_{j=m+1}^{\infty} \mathrm{var}(X_j)/s_j^2 \leq \sum_{j=m+1}^{\infty} P(A_j)/s_j^2$$

$$\leq \sum_{j=m+1}^{\infty} \int_{s_{j-1}}^{s_j} x^{-2}dx = \int_{s_m}^{\infty} x^{-2}dx = 1/s_m < \infty.$$

Also, by the same arguments,

$$\sum_{n=m}^{\infty} q(n) \sum_{j=n+1}^{\infty} c_j/s_j^2 \leq 2 \sum_{n=m}^{\infty} q(n)/s_n < \infty.$$

Thus Conditions (e) and (f) hold.   □

The proof of Theorem 4.2.2 shows that the following result is true: If $\{X_n\}_{n\geq1}$ are nonnegative and uniformly bounded sequence of random variables satisfying

$$\text{cov}(X_i, X_j) \leq q(j - i)EX_j \;\forall\, i > j,$$

and $\sum EX_n = \infty, \sum q(m)/E(S_m) < \infty$ where $q(n) \geq 0 \;\forall n \geq 1$, then $S_n/E(S_n) \to 1$ a.s. where $S_n = X_1 + \cdots + X_n, n \geq 1$.

We now state the following useful consequences of the above theorem.

Corollary 1 *Let $\sum P(A_n) = \infty$, and (4.2.1) hold where $\{q(n)\}_{n\geq1}$ is a nonnegative non-increasing sequence of reals.*

(a) If $\displaystyle\sum_{n=1}^{\infty} q(n) < \infty$, then $N_n/s_n \to 1$ a.s.;

(b) If $\limsup(n^\alpha P(A_n)) > 0$ and $\displaystyle\sum_{n=1}^{\infty} q(n)\, n^{\alpha-1} < \infty$ for some $\alpha \in [0, 1)$, then

$$P(\limsup A_n) = 1;$$

(c) If $\limsup(n P(A_n)) > 0$ and $\displaystyle\sum_{n=1}^{\infty} q(n)\, \log n < \infty$ then $P(\limsup A_n) = 1$.

Proof Part (a) follows, trivially, as $\displaystyle\sum_{n=1}^{\infty} q(n)/s_n \leq \sum_{n=1}^{\infty} q(n)/s_1 < \infty$.

For (b), get a subsequence $\{n_k\}_{k\geq1}$ of natural numbers and $\epsilon > 0$ such that

$$P(A_m) > \epsilon m^{-\alpha}, \quad \forall\, m = n_1, n_2, \ldots$$

Set $B_k = A_{n_k}, k \geq 1$. It suffices to show that $P(\limsup B_k) = 1$. Note that for all $i < j$,

$$P(B_i \cap B_j) - P(B_i)P(B_j) \leq q(n_j - n_i)P(B_j) \leq q(j - i)P(B_j);$$

here we have used the fact that $\{q(n)\}_{n\geq1}$ is nonincreasing. Also,

$$\sum_{j=1}^{k} P(B_j) \geq \delta\, k^{1-\alpha} \text{ for some } \delta > 0.$$

Thus,

$$\sum_{m=1}^{\infty} \left(q(m)/\sum_{j=1}^{m} P(B_j) \right) \leq \delta^{-1} \sum_{m=1}^{\infty} q(m)m^{\alpha-1} < \infty.$$

Hence, applying Theorem 4.2.2 to $\{B_k\}_{k\geq1}$, we get the desired result.

The proof of Part (c) is similar; see, e.g., page 70 of Chandra and Ghosal (1998). $\qquad\square$

Example 4.2.2 Let $\{X_n\}_{n\geq1}$ be a sequence of random variables such that $\exists$ a sequence $\{q(m)\}_{m\geq1}$ of non-negative reals satisfying $\sum_{m=1}^{\infty} q(m) < \infty$ and

$$P(X_i > s, X_j > t) - P(X_i > s)P(X_j > t) \leq q(j - i)P(X_j > t),$$

$$P(X_i < s, X_j < t) - P(X_i < s)P(X_j < t) \leq q(j - i)P(X_j < t)$$

for all $i < j, s, t \in \mathbb{R}$. Then for any real a, $P(X_n \to a) = 0$ or 1.

(If the X_n are pairwise m-dependent or pairwise NQD, then the above conditions are, trivially, satisfied.)

Solution: It suffices to show that

$$P(\limsup X_n \leq a) = 0 \text{ or } 1, \ P(\liminf X_n \geq a) = 0 \text{ or } 1.$$

To prove the first part, observe that

$$\left[\limsup X_n > a\right] = \bigcap_{m=1}^{\infty} [X_n > a + 1/m \ i.o. \ (n)].$$

If $\sum_{n=1}^{\infty} P(X_n > a + 1/m) < \infty \ \forall m \geq 1$, then $P(\limsup X_n \leq a) = 1$.

If $\exists$ an integer $m \geq 1$ such that $\sum_{n=1}^{\infty} P(X_n > a + 1/m) = \infty$, then $P(\limsup X_n \leq a) = 0$, as the events $\{[X_n > a + 1/m]\}_{n\geq1}$ satisfy the hypotheses of Corollary 1(a) of Theorem 4.2.2.

The second part can be proved from the first part applied to $\{-X_n\}_{n\geq1}$ and $-a$.

Remark 4.2.2 Assume that $\{X_n\}_{n\geq1}$ satisfies the conditions of the last example. Then $P(\{X_n\} \text{ converges to a finite limit}) = 0$ or 1; see Example 1.6.16.

A similar remark holds for Example 1.6.15.

References

T.K. Chandra, *Laws of Large Numbers* (Narosa Publishing House Pvt. Ltd., New Delhi, 2012)

T.K. Chandra, S. Ghosal, in *Some Elementary Strong Laws of Large Numbers: a review*, ed by S.P. Mukherjee, A.K. Basu, B.K. Sinha. Frontiers of Probability and Statistics, (Narosa publishing House Pvt, New Delhi, 1998)

T.K. Chandra, A. Goswami, Cesáro uniform integrability and strong laws of large numbers. Sankhyā, Ser. A, **54**, 215–231. Correction: Sankhyā, Ser. A, **55**, 327–328 (1992)

S. Csörgö, K. Tandori, V. Totik, On the strong law of large numbers for pairwise independent random variables. Acta Math. Hung. **42**, 319–330 (1983)

R. Durrett, *Probability: theory and examples, 3rd edn.* (Brooks/Cole, Belmont, 2005)

N. Etemadi, Stability of sums of weighted nonnegative random variables. JMA **13**, 361–365 (1983)

V.V. Petrov, On the strong law of large numbers for nonnegative random variables. TPA **53**, 346–349 (2008)

V.V. Petrov, On stability of sums of nonnegative random variables. JMS **159**, 324–326 (2009)

Chapter 5
The Conditional BCL

5.1 Lévy's Result

The assumption of 'independence of the A_n' in the second Borel–Cantelli lemma was replaced by weaker assumptions of some 'dependence' structures on A_n's in Chap. 3. An alternative way to get around this assumption is to use conditioning, which is, essentially, due to P. Lévy; see Lévy (1937, Corollary 68, p. 249) or Doob (1953, Corollary 2, p. 324).

To understand this form of Borel–Cantelli lemma, it is necessary to know the theory of conditional expectations and probabilities, and the theory of martingales; see, e.g., Chap. 6 of Billingsley (1995). In particular, any conditional probability will be a random variable, and not a numerical constant.

We shall use the following result about martingales; see, e.g., Theorem 5.2.8 on page 96 of Breiman (1968).

Theorem 5.1.1 *Let $\{(X_n, \mathcal{F}_n)\}_{n \geq 1}$ be a martingale such that*

$$E\left(\sup_{n \geq 1} |X_{n+1} - X_n|\right) < \infty.$$

If

$$A = \{\lim X_n \ \text{exists and is finite}\ ],$$

$$B = [\limsup X_n = \infty, \liminf X_n = -\infty],$$

then $P(A \cup B) = 1$.

(The above theorem resembles a *zero-one law*.)

T. K. Chandra, *The Borel–Cantelli Lemma*, SpringerBriefs in Statistics,
DOI: 10.1007/978-81-322-0677-4_5, © The Author(s) 2012

Theorem 5.1.2 (The Conditional Borel–Cantelli Lemma) *Let* $(\Omega, \mathcal{A}, P)$ *be a probability space, and let* $\{\mathcal{F}_n\}_{n\geq 1}$ *be an increasing sequence of sub-σ-fields of* $\mathcal{A}$, $\mathcal{F}_0 \subset \mathcal{F}_1 \subset \ldots$ *where* $\mathcal{F}_0 = \{\emptyset, \Omega\}$. *Then with probability one*

$$\sum_n 1_{A_n} = \infty \text{ iff } \sum_n P(A_{n+1}|\mathcal{F}_n) = \infty.$$

Proof Let $Z_n = I_{A_n} - P(A_n|\mathcal{F}_{n-1})$, $n \geq 1$. Then $|Z_n| \leq 1$ *a.s.* and $\{Z_n\}_{n\geq 1}$ is a martingale-difference sequence so that

$$X_n = \sum_{i=1}^{n} Z_i, \quad n \geq 1$$

is a martingale satisfying the condition of Theorem 5.1.1. Define A and B as in Theorem 5.1.1. Clearly, if $\omega \in A$ then

$$\sum_{n=1}^{\infty} I_{A_n}(\omega) = \infty \text{ iff } \sum_{n=1}^{\infty} P(A_{n+1}|\mathcal{F}_n)(\omega) = \infty.$$

If $\omega \in B$, then $\sum_{n=1}^{\infty} I_{A_n}(\omega) = \infty$ and $\sum_{n=1}^{\infty} P(A_{n+1}|\mathcal{F}_n)(\omega) = \infty$; use the facts that

$$\sum_{i=1}^{n} I_{A_i} \geq X_n, \quad \sum_{i=1}^{n} P(A_i|\mathcal{F}_{i-1}) \geq -X_n.$$

Since $P(A \cup B) = 1$ by Theorem 5.1.1, the proof is complete.   □

Lévy's theorem implies the usual Borel–Cantelli lemma. For, if $\sum P(A_n) < \infty$ then $E(\sum P(A_n|\mathcal{F}_{n-1})) < \infty$ and so $P(\sum P(A_n|\mathcal{F}_{n-1}) < \infty) = 1$ which implies, in view of Lévy's theorem, that $P(\sum 1_{A_n} < \infty) = 1$, i.e., that $P(\limsup A_n) = 0$; if $\{A_n\}_{n\geq 1}$ are independent and $\sum P(A_n) = \infty$, then taking $\mathcal{F}_0 = \{\emptyset, \Omega\}$ and $\mathcal{F}_n = \sigma(\{A_1, \ldots, A_n\})$ for $n \geq 1$ we get $\sum P(A_n|\mathcal{F}_{n-1}) = \sum P(A_n)$ *a.s.* so that $P(\sum 1_{A_n} = \infty) = 1$, i.e., $P(\limsup A_n) = 1$. Lévy's result is, however, more widely applicable, as the following example demonstrates.

Example 5.1.1 Let $\{X_n\}_{n\geq 1}$ be iid with the common distribution $U(0, 1)$, the uniform distribution over $(0, 1)$. Then $X_1, \ldots, X_n$ splits $(0, 1)$ into $(n + 1)$ subintervals. Let $A_n = [X_{n+1} \in I_n]$ where I_n is the largest of these subintervals. Show that $P(\limsup A_n) = 1$.

Solution: Clearly, the length of I_n is $\geq 1/(n + 1)$, and

$$P(A_n|\mathcal{F}_{n-1}) = \text{ the length of } I_n \geq 1/(n + 1) \text{ a.s.}$$

where $\mathcal{F}_n = \sigma(\{X_1, \ldots, X_n\}), n \geq 1$. Thus, $\sum P(A_n | \mathcal{F}_{n-1}) = \infty$ a.s. and hence $\sum 1_{A_n} = \infty$ a.s. by Lévy's theorem. So $P(\limsup A_n) = 1$. $\qquad \square$

We now give a very simple and elementary proof, due to Chen (1978), of a slightly more general form of Theorem 5.1.2 (see, also, Meyer (1972) and Freedman (1973)).

Theorem 5.1.3 (Chen (1978)) *Let $\{X_n\}_{n\geq 1}$ be a sequence of non-negative random variables defined on $(\Omega, \mathcal{A}, P)$. Let $\{\mathcal{F}_n\}_{n\geq 0}$ be a sequence of sub-σ-fields of $\mathcal{A}$. Let $M_n = E(X_n | \mathcal{F}_{n-1})$ for $n \geq 1$.*
If $\{\mathcal{F}_n\}_{n\geq 0}$ is increasing, i.e., if $\mathcal{F}_n \subset \mathcal{F}_{n+1} \forall n \geq 0$, then

$$\sum_{n=1}^{\infty} X_n < \infty \ a.s. \ on \ \left[\sum_{n=1}^{\infty} M_n < \infty\right].$$

Conversely, if $Y := \sup_{n\geq 1}(X_n/(1+X_1+\cdots+X_{n-1}))$ is integrable and $\sigma(X_1+\cdots+X_n) \subset \mathcal{F}_n \forall n \geq 1$, then

$$\sum_{n=1}^{\infty} M_n < \infty \ a.s. \ on \ \left[\sum_{n=1}^{\infty} X_n < \infty\right].$$

(X_n need not be, in general, $\mathcal{F}_n$-measurable.)

Proof Let $M_0 \equiv 1$. Note that

$$\sum_{n=1}^{\infty} M_n/(S_{n-1}S_n) \leq 1$$

where $S_n = \sum_{i=0}^{n} M_i, n \geq 0$; this is true, since the above series is telescoping and its

sum is $1 - \left(\sum_{n=0}^{\infty} M_n\right)^{-1}$. Therefore,

$$1 \geq E\left(\sum_{n=1}^{\infty} M_n/(S_{n-1}S_n)\right)$$
$$= E\left(\sum_{n=1}^{\infty} X_n/(S_{n-1}S_n)\right) \geq E\left(\left(\sum_{n=1}^{\infty} X_n\right) \bigg/ \left(\sum_{n=0}^{\infty} M_n\right)^2\right).$$

[Here we have used the Monotone Convergence Theorem (applied to a nonnegative series), and the fact that

$$E\left(X_n/(S_{n-1}S_n)\right) = E(E(X_n/(S_{n-1}S_n)|\mathcal{F}_{n-1}))$$

$$= E(M_n/(S_{n-1}S_n)).]$$

Therefore, $P\left(\left[\sum_{n=1}^{\infty} X_n = \infty\right] \cap \left[\sum_{n=1}^{\infty} M_n < \infty\right]\right) = 0.$

To prove the converse, let $X_0 \equiv 1$. Note that, if $S_n^* = \sum_{i=0}^{n} X_i$ for $n \geq 0$,

$$E\left(\left(\sum_{n=1}^{\infty} M_n\right)\Big/\left(\sum_{n=0}^{\infty} X_n\right)^2\right) \leq E\left(\sum_{n=1}^{\infty} M_n/S_{n-1}^{*2}\right)$$

$$= E\left(\sum_{n=1}^{\infty} X_n/S_{n-1}^{*2}\right)$$

$$= E\left(\sum_{n=1}^{\infty}(X_n/(S_{n-1}^* S_n^*))(1 + X_n/S_{n-1}^*)\right)$$

$$\leq E\left((1+Y)\sum_{n=1}^{\infty} X_n/(S_{n-1}^* S_n^*)\right)$$

$$\leq E(1+Y) < \infty.$$

This implies that $P\left(\left[\sum_{n=1}^{\infty} M_n = \infty\right] \cap \left[\sum_{n=1}^{\infty} X_n < \infty\right]\right) = 0.$ □

(See the last paragraph on page 700 of Chen (1978) for some useful remarks.)

Corollary 1 *Let $\{X_n\}_{n\geq 1}$ be a sequence of nonnegative random variables defined on $(\Omega, \mathcal{A}, P)$ and let there be an increasing sequence $\{\mathcal{F}_n\}_{n\geq 0}$ of sub-σ-fields of $\mathcal{A}$. Let $\mathcal{G}_0 \subset \mathcal{A}$ be any σ-field, and for $n \geq 1$, let $\mathcal{G}_n = \sigma(X_1 + \ldots + X_n)$. Suppose that $E(Y) < \infty$ where Y is as in Theorem 5.1.3. Then*

$$\left[\sum_{n=1}^{\infty} E(X_n|\mathcal{F}_{n-1}) < \infty\right] \subset_P \left[\sum_{n=1}^{\infty} E(X_n|\mathcal{G}_{n-1}) < \infty\right]$$

where $A \subset_P B$ means that $P(A \cap B^c) = 0$.

Proof This is immediate from the above theorem:

$$\left[\sum_{n=1}^{\infty} E(X_n|\mathcal{F}_{n-1}) < \infty\right] \subset_P \left[\sum X_n < \infty\right] \subset_P \left[\sum_{n=1}^{\infty} E(X_n|\mathcal{G}_{n-1}) < \infty\right]. \quad □$$

We next state a result of Dubins and Freedman (1965).

Theorem 5.1.4 *Let $(\Omega, \mathcal{A}, P)$ be a probability space. Let $\{A_n\}_{n\geq 1}$ be a sequence of events such that $A_n \in \mathcal{F}_n \; \forall n \geq 1$ where $\{\mathcal{F}_n\}_{n\geq 0}$ is an increasing sequence of sub-σ-fields of $\mathcal{A}$. Let $p_n = P(A_n|\mathcal{F}_{n-1}), n \geq 1$ and assume that $0 < p_1 < 1, \mathcal{F}_0 = \{\emptyset, \Omega\}$. Then*

$$(I_{A_1} + \cdots + I_{A_n})/(p_1 + \cdots + p_n)$$

converges to a finite limit L a.s. and in rth mean $(0 < r < \infty)$. Also, $L = 1$ a.s. on

$$\left[\sum_1^\infty p_n = \infty \right].$$

A part of this theorem is proved in Athreya and Lahiri (2006, pp. 235–236). For generalizations, see Freedman (1973, Proposition 39 and Theorem 40 on pages 920 and 921 respectively; see, also, Proposition 52 on page 925).

5.2 A Result of Serfling

We first introduce a notation. For random variables X and Y, let

$$d(X, Y) = \sup_{B \in \mathcal{B}} |P(X \in B) - P(Y \in B)|$$

where $\mathcal{B}$ is the Borel σ-field on $\mathbb{R}$. Then

(a) $d(X, Y) \leq P(X \neq Y)$, provided X, Y are defined on the same probability space, and

(b) $d(X, Z) \leq d(X, Y) + d(Y, Z)$.

Thus, $d(X, Y)$ can be regarded as a distance, called the *total variation distance*, between X and Y (or, more precisely, between the distributions of X and Y).

Results of this section are due to Serfling (1975).

Lemma 5.2.1 *Let $X_1, \ldots, X_n$ be non-negative integer-valued random variables defined on a probability space $(\Omega, \mathcal{A}, P)$. Put*

$$p_1 = P(X_1 = 1), \quad p_i = P(X_i = 1|\mathcal{F}_{i-1}) \text{ for } 2 \leq i \leq n,$$

where $\mathcal{F}_i = \sigma(X_1, \ldots, X_i)$. Let $X_1^*, \ldots, X_n^*$ be independent Bernoulli variables with respective success probabilities $p_1^*, \ldots, p_n^*$. Then

$$d\left(\sum_{i=1}^n X_i, \sum_{i=1}^n X_i^* \right) \leq \sum_{i=1}^n E|p_i - p_i^*| + \sum_{i=1}^n P(X_i \geq 2).$$

Proof Write $X'_i = I_{[X_i=1]}$ for $1 \le i \le n$, and put

$$p'_1 = p_1, \ p'_i = P(X'_i = 1|\mathcal{F}'_{i-1}) \text{ for } 2 \le i \le n$$

where $\mathcal{F}'_i = \sigma(X'_1, \ldots, X'_i)$. Clearly, $\mathcal{F}'_i \subset \mathcal{F}_i$ and so

$$p'_i = E(p_i|\mathcal{F}'_{i-1})) \text{ for } 2 \le i \le n.$$

Now note that

$$d\left(\sum_{i=1}^{n} X_i, \sum_{i=1}^{n} X'_i\right) \le P\left(\sum_{i=1}^{n} X_i \ne \sum_{i=1}^{n} X'_i\right)$$

$$\le \sum_{i=1}^{n} P(X_i \ne X'_i) = \sum_{i=1}^{n} P(X_i \ge 2).$$

We next proceed to construct X'_i and X^*_i for $1 \le i \le n$ on a common probability space. Let $R_1, \ldots, R_n$ be iid random variables following the uniform distribution on [0,1]. Set

$$X^*_i = I_{[R_i \le p^*_i]} \text{ for } 1 \le i \le n, \text{ and}$$

$$X'_1 = I_{[R_1 \le p'_1]}, \quad X'_i = I_{[R_i \le p'_i(X'_1,\ldots,X'_{i-1})]}, \quad 2 \le i \le n$$

where, for $2 \le i \le n$,

$$p'_i(x'_1, \ldots, x'_{i-1}) = P(x'_i = 1|x'_1 = x'_1, \ldots, x'_{i-1} = x'_{i-1}),$$

$$x'_j = 0 \text{ or } 1, \ 1 \le j \le i - 1.$$

Now observe that

$$P(X'_i \ne X^*_i) = E(P(X'_i \ne X^*_i|\mathcal{F}'_{i-1})) \le E(|p'_i - p^*_i|)$$

and that

$$E(|p'_i - p^*_i|) = E(|(p_i - p^*_i|\mathcal{F}'_{i-1})|) \le E(|p_i - p^*_i|)$$

Thus $d\left(\sum_{i=1}^{n} X'_i, \sum_{i=1}^{n} X^*_i\right) \le \sum_{i=1}^{n} E(|p_i - p^*_i|).$ $\qquad\square$

The following result is due to Iosifescu and Theodorescu (1969, p. 2): let $\phi_1 = 0$, and for $n \ge 2$

$$\phi_n = \sup\{|P(A_n|F) - P(A_n)| : F \in \mathcal{F}_{n-1}, P(F) > 0\}$$

where $\mathcal{F}_n = \sigma(A_1, \ldots, A_n)$, $A_1, A_2, \ldots$ being a given sequence of events; if $\sum \phi_n < \infty$, then $P(A_n \ i.o.) = 1$. The next result, due to Serfling (1975), gives an extension of the above; see Remark 5.2.1 below.

Theorem 5.2.1 *Let $\{A_n\}_{n \geq 1}$ be a sequence of events. Put*

$$p_1 = P(A_1), \quad p_n = P(A_n | \mathcal{F}_{n-1}) \text{ for } n \geq 2.$$

If $\sum_{n=1}^{\infty} E(|p_n - P(A_n)|) < \infty$ and $\sum P(A_n) = \infty$, then $P(A_n \ i.o.) = 1$. Here $\mathcal{F}_n = \sigma(\{A_1, \ldots, A_n\})$, $n \geq 1$.

Proof It suffices to show that $P\left(\bigcap_{n=m}^{\infty} A_n^c\right) \to 0$ as $m \to \infty$. To this end, let $X_n = I_{A_n}$, $n \geq 1$. Then $\mathcal{F}_1, \mathcal{F}_2, \ldots$ and $p_1, p_2, \ldots$ as defined above also correspond to $X_1, X_2, \ldots$ as in Lemma 5.2.1. Hence,

$$P\left(\bigcap_{n=m}^{M} A_n^c\right) = P\left(\sum_{n=m}^{M} X_n = 0\right)$$

$$\leq P\left(\sum_{n=m}^{M} X_n^* = 0\right) + d\left(\sum_{n=m}^{M} X_n, \sum_{n=m}^{M} X_n^*\right)$$

$$\leq P\left(\sum_{n=m}^{M} X_n^* = 0\right) + \sum_{n=m}^{M} E\left(|p_n - p_n^*|\right)$$

where $X_1^*, X_2^*, \ldots$ are independent Bernoulli variables with respective success probabilities $P(A_1) = p_1^*$, $P(A_2) = p_2^*, \ldots$ Using the independence,

$$P\left(\sum_{n=m}^{M} X_n^* = 0\right) = \prod_{n=m}^{M} (1 - P(A_n)) \leq \exp\left(-\sum_{n=m}^{M} P(A_n)\right).$$

Letting $M \to \infty$, we get in view of $\sum P(A_n) = \infty$

$$P\left(\bigcap_{n=m}^{\infty} A_n^c\right) \leq \sum_{n=m}^{\infty} E(|p_n - P(A_n)|).$$

Letting $m \to \infty$, we get the desired result. $\qquad \square$

Remark 5.2.1 Note that $|p_n - P(A_n)| \leq \phi_n$ a.s. This can be seen in the following way: let $n \geq 2$, and note that $\mathcal{F}_{n-1}$ is atomic; let the atoms of it having non-zero probabilities be $E_1, \ldots, E_k$. Then $\sum_{j=1}^{k} I_{E_j} = 1$ a.s., and

$$P(A_n | \mathcal{F}_{n-1}) = \sum_{j=1}^{k} I_{E_j} P(A_n | E_j) \ a.s.$$

so that

$$|p_n - P(A_n)| = |\sum_{j=1}^{k} (P(A_n | E_j) - P(A_n)) I_{E_j}| \ a.s.$$

$$\leq \sum_{j=1}^{k} I_{E_j} \phi_n = \phi_n \ a.s.$$

References

K.B. Athreya, S.N. Lahiri, *Probability Theory*, Trim Series 41. (Hindustan Book Agency, India 2006)

P. Billingsley, *Probability and Measure*, 3rd edn. (Wiley, New York, 1995). Second Edition 1991. First Edition 1986

L. Breiman, *Probability* (Addision Wesley, California, 1968)

L.H.Y. Chen, A short note on the conditional Borel–Cantelli lemma. AP **6**, 699–700 (1978)

J.L. Doob, *Stochastic Processes* (Wiley, New York, 1953)

L.E. Dubins, D.A. Freedman, A sharper form of the Borel–Cantelli lemma and the strong law. AMS **36**, 800–807 (1965)

D. Freedman, Another note on the Borel–Cantelli lemma and the strong law, with the Poisson approximation as a by-product. AP **1**, 910–925 (1973)

M. Iosifescu, R. Theodorescu, *Random Processes and Learning* (Springer, New York, 1969)

P. Lévy, *Théorie de l' addition des variables aléatoires* (Gauthier-Villars, Paris, 1937)

P.A. Meyer, *Martingales and Stochastic Integrals I*. Lecture Notes in Math. 284, (Springer, Berlin, 1972)

R.J. Serfling, A general poisson approxiation theorem. AP **3**, 726–731 (1975)

Author Index

T. K. Chandra, *The Borel–Cantelli Lemma*, SpringerBriefs in Statistics,
DOI: 10.1007/978-81-322-0677-4, © The Author(s) 2012

Subject Index

T. K. Chandra, *The Borel–Cantelli Lemma*, SpringerBriefs in Statistics,
DOI: 10.1007/978-81-322-0677-4, © The Author(s) 2012